MANAGING SNOW&ICE

BY JOHN A. ALLIN

EDITOR
Steve Smith

ART DIRECTOR
Karen Angus

PRODUCTION MANAGER
Helen Duerr

TECHNICAL RESOURCE MANAGER
Anita Tucker

PROOFREADERS
Lisa Edwards, Darlene Taylor

PUBLISHER: BOB WEST
GIE MEDIA

Address all correspondence to GIE Media, 4012 Bridge Ave., Cleveland, OH 44113, or call 216/961-4130.

Library of Congress Control Number: 2002141405
ISBN: 1-883751-17-9

CONTENTS

ABOUT THE AUTHOR

John Allin is president of Allin Companies and Snow Management Group, Erie, Pa., and one of the foremost snow and ice industry experts. He has been involved in the snowplowing industry for more than 20 years and is a founding member of the Snow & Ice Management Association (SIMA), a trade association of private snow removal contractors. From 1996 to June 2002, Allin served as SIMA board president.

In 1978 after attending Gannon University, Erie, Pa., Allin started the company with one truck and a new snowplow. Today, Snow Management Group has grown to include more than 1,750 subcontractors, 6,000 plowing and salting units operating in 15 states. Snow and ice management revenues for the 2002-2003 winter season are projected to be in excess of $40 million. Allin's snow removal customers include many large commercial and industrial accounts, as well as smaller retail establishments. Erie, Pa., receives challenging lake-effect snow — often more than 250 inches in the snowbelts just inland from the shore of Lake Erie.

In 1998 Allin began providing consultation and winter management services to facility managers and contractors throughout the United States. Currently the company manages more than $20 million in snowplowing contracts east of the Mississippi River. Allin has consulted on snowplowing projects that include Newark International Airport, JFK International Airport, U.S. Air Force Academy, Lucent Technologies and The Gap. In 2002, Snow Management Group served as the snow removal contractor for the Winter Olympics in Salt Lake City, Utah.

Allin often serves as an expert witness on snow and ice issues and is qualified as an "expert" in seven states.

ACKNOWLEDGEMENTS

This book is the culmination of years of dedication and hard work. I am guided by a few very valuable lessons, which have allowed me to succeed in my chosen profession.

My high school hockey coach (Jack Moran) did not pick me to be MVP of our team my senior year. Afterward he told me I was indeed the most valuable player, but he gave the award to another person who had been on the team longer. That incident taught me life is not fair, and I have never forgotten it.

My editor (Steve Smith) has guided me through the writing of this book and has been patient with me throughout this past year. He pushed and prodded me when it was right to do so and gave me praise when it was needed most. My publisher (Bob West) gave me the confidence to proceed with this challenge. The late Phil Christian gave me the confidence to think outside the box and to dream impossible dreams. He would have been proud to see what I have accomplished and to know he had a hand in it. He is missed.

Most of all, my wife (Peggy) has allowed me to follow my dreams when others advised me to select the safe and easy route. Her love and support has made it possible for me to do things in this industry others thought impossible. None of the things I have accomplished in the past 15 years would have been possible without her.

John A. Allin
June 2002

DEDICATION

My father (Whitey Allin) taught me that if I am convinced I can do something, then I can. From that lesson I developed my mantra, "If you think you can, or you think you can't, you're right." My father has influenced my life profoundly and continues to be a sounding board I find invaluable. I dread the day I shall not be able to talk through my problems with him. I have never said "I love you" enough. I love you, Dad. Thank you.

John A. Allin
June 2002

PREFACE

Snow and ice management is a multi-billion dollar business. And yet, those business professionals who provide this necessary and valuable service are considered by some to be "plow jockeys." The connotation is that we cannot find "real jobs," so that is why we are out all night, working in horrendous conditions doing a job that supposedly "anybody can do."

As we in the business know, this is not the case. This industry takes significant technical ability, keen business sense, perseverance and dedication.

Unfortunately, there are some snowplow operators who don't know anything about how to run a business. Phrases and words such as capital budgets, operating budgets, cash flow projections, balance sheets, financial statements and sales forecasts are foreign to an unacceptably large portion of contractors. Laws on the use of subcontractors, human resources, contracts and tax issues also present further complication. Many of these smaller operators want to grow their one-truck operations into large, money-making companies, but they don't know what is required or necessary in order to grow.

This book is an effort to break down the various components needed to successfully operate a snow and ice removal business. From how to operate a snowplow to how to price services, this text is designed to educate someone who has a desire to be a snow business professional.

I have been fortunate to visit hundreds of contractor operations in all parts of the United States and Canada. I have talked with thousands of plowing contractors of various sizes and levels of expertise. Through my association with the Snow & Ice Management Association, I have had the privilege of watching our industry grow from a fractured assortment of people trying to do well, to an organized group of professionals making a difference. From this interaction I have been fortunate to meet some incredibly intelligent individuals who are quite successful at this business. I have gained innumerable insights into how businesses operate, what pitfalls to avoid and how to be profitable.

I have also been fortunate to live in an area of the country that receives considerable lake-effect snow. This has allowed me to develop a management system that is smooth and effortless with regard to the most efficient ways to move snow and manage ice on customer locations. And, make no mistake, we are doing just that, managing the property.

That is what makes the snow removal business so challenging and rewarding — managing the snowstorm and not letting the storm manage you. The snow business is a crisis-management industry. We must systematically and methodically address each crisis if we are to make a profit and keep our customers happy. My sincere hope is that this text will help you achieve that goal.

John A. Allin
June 2002

PART ONE

BUSINESS MANAGEMENT

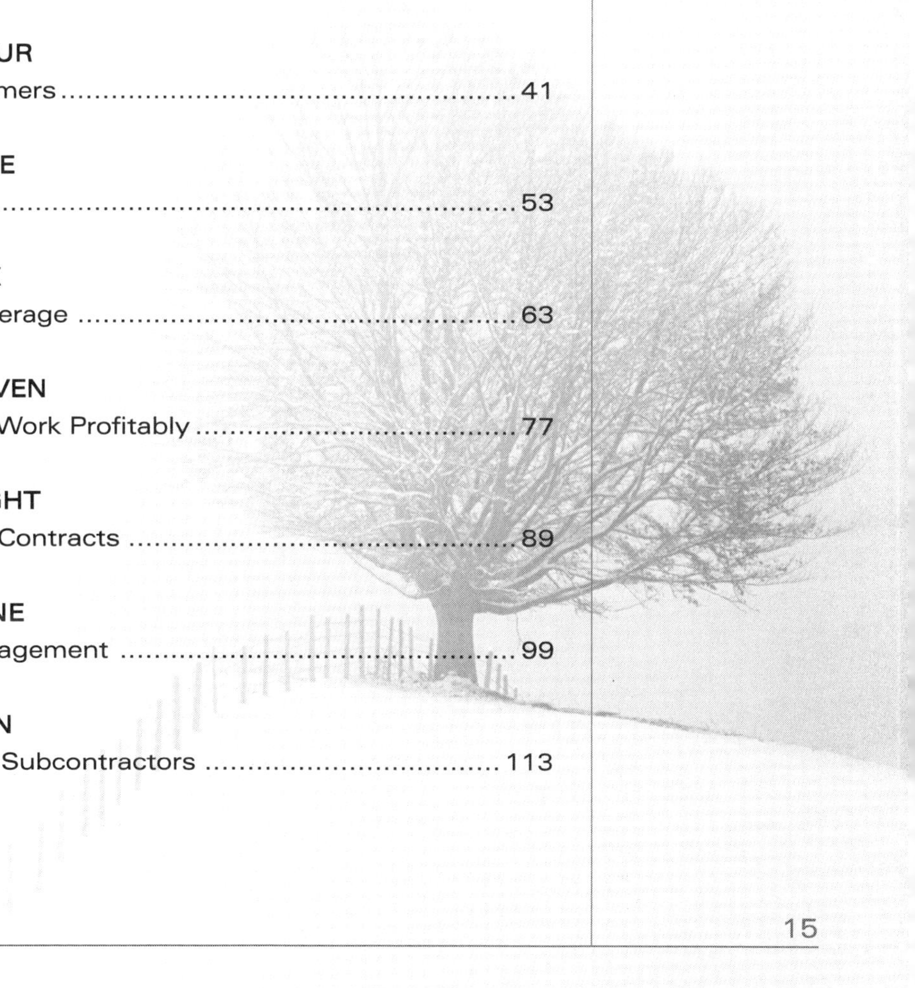

CHAPTER ONE: A History Of Snow Removal

IN THIS CHAPTER

- ❄ Early methods of snow and ice management
- ❄ The modernization of snow removal
- ❄ The creation of the private professional snow removal industry

As families across the city made their final preparations for the Christmas holiday, snowflakes started to descend from the skies over Buffalo, N.Y. Falling heavier by the minute, it soon became apparent this was no ordinary snowstorm. More than 2 feet of snow hammered Buffalo on Dec. 24, 2001 effectively shutting down the city and Buffalo Niagara International Airport. Three days later another 25.6 inches of lake-effect snow piled up, creating a total of more than four feet between Dec. 24 and 27. The 29.8 inches of snow that fell from Wednesday, Dec. 26, through Thursday, Dec. 27, was the second snowiest 24-hour period in Buffalo.

In total, nearly 7 feet of lake-effect snow fell on the city between Christmas Eve and New Year's Day, prompting U.S. President George W. Bush to declare a state of emergency for areas of upstate New York and the city of Buffalo.

The storm was brutal, causing many injuries and countless economic losses due to temporary business closings. But the storm's damaging impact would have been much worse, if not for the efforts of snow and ice removal professionals, who helped dig out the city inch-by-inch. After a few days of intense work, life slowly returned to normal in Buffalo.

Responding to extreme snow events is a part of daily life in snow and ice management. Note the cars buried in snow.

Stories like this, although on a smaller scale, happen every winter across much of the north-

SNOW FACT: The heaviest 24-hour snowfall on record in the continental United States happened in Silver Lake, Colo., when 76 inches fell April 14-15, 1921.[1]

ern half of the U.S. and throughout Canada. Snow and ice professionals work under harsh conditions, for long hours, to fight Mother Nature in her own backyard. Commerce, transportation and life as we know it would not proceed in the winter without these dedicated professionals.

The work isn't completed out of sheer selflessness, however. Snow and ice removal, if managed properly, can be very profitable. Indeed, a few in the industry make their entire year's revenue from just a few months each year of snow removal.

EARLY SNOW CONTROL

Those dedicated snow and ice professionals across North America are part of a long line of individuals who, since the modern age began, have been finding ways to deal with winter precipitation. Acording to *Have Snow Shovel, Will Travel* (National Snow and Ice Data Center), the earliest attempts at snow control in North America involved leveling snow drifts for sleigh traffic during the 1700s. Many city laws required business owners to clear snow from the sidewalks and streets in their area. Wintertime travel in the early 1800s was mostly conducted on foot.

The 19th century brought increased industrialization and commerce to North America, making snow removal more important. In the 1840s, inventors designed and patented the first snowplows, but several years passed before the plows were widely used. Reportedly, one of the first mentions of snowplow use comes from Milwaukee in 1862. The plow was attached to a cart pulled by a team of horses.

Before development of modern snow-fighting techniques, winter storms could paralyze travel and communication for long periods of time.

Throughout the next several years, horse-drawn plows gained popularity and came into use in many other northeastern cities. Soon after the first plows were implemented on a wide scale, retail businessmen began complaining, claiming that their storefronts were completely blocked with mounds of snow, making them inaccessible to customers. Sleigh drivers also found fault with the plowing system because of the ruts and uneven surfaces it created.

Cities along the eastern U.S. and in the Snowbelt responded in several ways. They hired horse-drawn carts and individuals with shovels to work in conjunction with the plows, hauling away the plowed snow and dumping it into rivers (a definite "no-no" in today's environmentally conscious world).

THE BIG ONE

A major push for advancement in snow removal techniques came as a result of the notorious "Blizzard of 1888," which paralyzed the Northeast after three days of snow, wind and freezing temperatures. Reportedly, snow covered the entire first levels of buildings, as more than 2 to 4 feet of snow fell and drifted.

The storm took the lives of more than 400 people and immobilized fire departments, passenger trains and communication across much of the Northeast. New York, Boston, Philadelphia and Washington, D.C., were cut off from communication with one another for several days when telegraph and telephone wires were downed by high winds and heavy accumulations of ice and snow.

Some early motorized equipment was developed to deal with difficult snow and ice conditions.

Following the 1888 blizzard, cities recognized the need for more organized snow removal and were motivated to bury communication wires and create alternative methods of transportation, such as trains and subways, which wouldn't be hindered by drifting snow.

HERE COME THE CARS

Once the automobile became popular, an increasing demand was placed on removing snow from city streets. Likewise, cities began "motorizing" their snowplow fleets and began using dump trucks, tractors and heavy equipment for snow removal.

In 1920, the Barber-Green snow loader was invented and

The advent of satellites drastically improved weather forecasting.

successfully introduced in Chicago. Several cities purchased the machine thereafter. The snow loader used tractor treads, a scoop and conveyor belt. As the snow was plowed, it was forced up the scoop, caught by the conveyor belt and carried up and away from the street into a chute at the top. The snow was dropped into a dump truck parked underneath.

Around the same time of the Barber-Green snow loader, financial pressure from the loss of business due to snow and ice storms and the resulting travel difficulties prompted cities to invest in new fleets of dump trucks and tractor plows.

Ice left behind by snowplowing called for increased use of salt and sand, and with the need for clear streets outweighing environmental concerns, city public works officials used salt by the ton to ease road conditions, while also experimenting with cinders and sand.

Truck-mounted salt spreaders became the primary tool in fighting snowy roads and businesses as private citizens used tons of salt to keep driveways, sidewalks and access routes clear of snow and ice.

NOTABLE SNOW-STORM: February and March 1717: "The Great Snow of 1717" blanketed New England in a series of four storms, leaving nearly 4 feet on the ground and drifts as high as 25 feet.[2]

TECHNOLOGICAL BREAKTHROUGHS

The refinement of weather forecasting, including the use of satellites, began to assist snow professionals of the late 1950s and 1960s. As a result, snowplow operators could prepare in advance for storms.

As equipment became more efficient and effective, plow technology advanced too. Hydraulics were introduced in the 1960s, which allowed much more efficient operation of snowplows. Along with hydraulically operated plows came increased ease of operation. Hydraulics made it much easier to maneuver the plow setup and allowed operators to stay inside the cab to move the plow from side to side.

The addition of electrically operated switches enhanced operation, but the mechanics of moving snow from pavement surfaces remained basically the same. Loaders made it easier to move large quantities of snow but were very inefficient in that an operator was using a large piece of equipment to move a (relatively) light amount of product. Skid steers (small loaders) have made sidewalk snow clearing quicker, as these units are considerably more productive than a human operating by himself or herself.

Unfortunately, plowing snow is done today much like it was 75 years ago. Advances in the technology of actually moving snow have been minor and mostly limited to making the ease of operation more attractive. The advent of motorized vehicles made the process faster, but the process is still the same. Even today snowplowing operations are a reflection of the same philosophy in that we push snow out of the way, usually to the left or right of the vehicle.

AN INDUSTRY IS BORN

Long after automobile use became widespread, shopping centers, office parks and industrial centers saw the need for having their parking lots and pedestrian areas free of snow and ice. Many of these companies turned to private snow removal contractors or purchased equipment of their own.

Today, the private snow removal industry is a growing and well-established market. Though no hard numbers exist, it is believed that roughly 20,000 private companies across the U.S. perform snow and ice removal, generating approximately $6 billion in annual revenue. That's not to mention the countless numbers of subcontractors and "one-man" operations who also remove snow.

My experience in talking with successful plowing contractors throughout the U.S. and Canada indicates gross profit margins for snow removal in excess of 60 percent are normal, and gross profit margins for ice control services in excess of 70 percent are achievable. Those contractors who view snow removal as a profit center regularly state that snow is the most profitable part of their businesses. From contractors who project five snow events per season in a great winter to those who project 35 snow events in a mild winter, all report gross profits at or in excess of those reported above. Snowplowing is profitable, if priced and managed properly. ❄

SNOW TERM: THE SNOWBELT
Traditionally, the Northeast and Midwest play host to the most severe snowstorms. The area defined as the Snowbelt, stretching across the Great Lakes from Minnesota to Maine, usually receives the brunt of winter storms.

SOURCES/FURTHER READING

[1]National Weather Service.

[2]National Snow & Ice Data Center, University of Colorado, Boulder, Colo., http://nsidc.org, AccuWeather.com, www.accuweather.com, National Climatic Data Center, National Oceanic and Atmospheric Administration.

Have Snow Shovel, Will Travel by Laura Cheshire, Roger G. Barry, Annette Varani and Mike Meshek, National Snow & Ice Data Center, www.nsidc.org.

Snow In The Cities by Blake McKelvey, University of Rochester Press.

CHAPTER TWO: Getting Started

IN THIS CHAPTER

- ❄ Things to consider before entering the snow and ice field
- ❄ Equipment and materials you'll need to get started
- ❄ Working as a subcontractor
- ❄ Adding snow removal to your existing business

Entering the snow removal business requires dedication and perseverance. As with any business, a snow and ice management company needs to be managed effectively and profitably. Too often, snow is considered an ancillary service to another business. These core businesses often include landscape contracting, pavement maintenance, excavation and/or trucking. Sometimes the decision to enter the snow management business is dictated by attempts to increase cash flow during slower times of the year or by customers who demand the service.

All of the above are valid reasons to enter this industry. However, a more appropriate reason should be to develop a profit center that will augment the existing business. An increasing number of contractors have decided snow management *is* their core business.

WHO ARE SNOW PROFESSIONALS?

Landscape contractors seem to have a more natural avenue into the snow business, as often their customers require winter services. Such contractors have a good portion of the equipment already available. And the jump from two-wheel-drive vehicles to four-wheel-drive vehicles in the purchasing decision is not too difficult. Further, the landscape

Landscape contractors often have a natural avenue into the snow business.

contractor also has ready access to the personnel required for snow operations and has the me-

NOTABLE SNOWSTORM: January 1772: "The Washington and Jefferson Snowstorm" is so named because it trapped both men at their homes with snow as deep as 3 feet throughout Maryland and Virginia.[1] ❄

Business owners have much to consider before entering the snow industry.

chanical ability to fix inevitable equipment breakdowns.

Pavement installation and maintenance contractors, as well as excavation contractors, who usually have access to larger equipment, are often involved in larger commercial plowing accounts. When a melding of these contractors happens, there can be an awesome service capability attained, which can lead to serving the very largest of accounts.

THINGS TO CONSIDER

No matter your current business or your motivation for considering snow removal, there are many things to consider before jumping into snow removal.

- How much money do you invest?
- Do you use all of your own equipment, or use subcontractors?
- In growing this portion of your business, do you concentrate on the residential or commercial market?

Snow removal can often involve high stress and demanding customers.

High Stress. The snow business is usually a high margin business. However, sometimes all that money is clouded by sleep deprivation, angry spouses, disappointed children and unfeeling customers. Before you start, make sure you have seriously considered your level of commitment to the endeavor.

Demanding Customers. Snowplowing isn't for everyone. It takes long hours and a lot of sleepless nights. Customers can be very demanding and really don't care about how much (or how little) sleep you get. Snow falls at some very inopportune times and missing a family Hanukkah, Christmas, New Year's Eve or even Thanksgiving is a very real possibility. Often customers will call just as the first gift is being opened because they want to get to family gatherings or parties.

For example, a residential customer once called me at 8 a.m. on Christmas Day demanding that we plow his driveway so he could go to church services. Sixteen inches were already on the ground and no city plows, which would normally remove the snow from the city streets, had been dispatched. I plowed the driveway and then explained to the homeowner that he would only be able to make it to the end of the driveway.

After plowing out another resident (with much the same intention of going out to get a loaf of bread), I went by the first residence only to see a his car stuck at the entrance to the driveway. I got no satisfaction from being correct.

However, this sort of attitude on the part of the residential customer is commonplace, frustrating and often difficult to envision. This is part of the lives we choose when entering the snow and ice management business.

Commercial customers are no less demanding. Retail facilities are now open on most holidays and often demand that their parking lots and facilities be free of snow 24 hours a day, seven days a week. Regardless of your own personal feelings, the customer must come first in such situations if you hope to achieve customer satisfaction levels that are necessary to your business' success and growth.

WHAT YOU NEED TO GET STARTED

Some people enter the snow business with a snowblower and a snow shovel. This is the snow business at its most basic. Usually, this is a person who has yet to acquire a driver's license and may still be in high school or college. Though simple, this model can be profitable if addressed properly. Snow blowers cost from $350 to $1,500 depending upon the specifications. (See Chapter 11, Snow Removal Equipment for more detail.) Yet, it is difficult to grow the business to any significant level if you only have the snowblower or shovel at your disposal.

A straight-blade plow mounted on a pickup truck is by far the most common snow removal tool for residential and light commercial work.

Adding Vehicles. To begin growing, the need for a basic plowing vehicle becomes necessary. For an entry-level business, this usually involves a basic pickup truck mounted with a straight-blade snowplow.

All the major truck manufacturers have their good and bad points in relation to snow removal. I have owned most types and do not necessarily favor one over another. Specific manufacturers aside, there are some

Snow removal at its most basic: a pickup truck, snowblower, shovel and walk-behind salt spreader.

things that you should look for in selecting and ordering a pickup truck for plowing.

Four-wheel-drive trucks are usually recommended for plowing snow. Two-wheel-drive trucks can work if tire chains are used, or if the truck is large enough to hold enough weight to allow the drive wheels to have traction. (Consult Chapter 11, Snow Removal Equipment, for more detail.)

Plow Time. The next investment is acquiring a plow for your truck. The options are fairly limited, thus making the choice much easier. Straight blades and "V blades" are the most commonly used in the industry. Recently some variations of these blades have come on the market. These newer designs include plows that have hydraulically operated extensions on either end of the moldboard — the angled plate at the front of the plow — and even plows that flip over to create a back dragging pull plow on the front of the truck.

Numerous manufacturers build straight-blade plows, which come in several "duties" depending on the application. One should be very cautious not to purchase a plow that is too light-duty for the intended use. Saving a few dollars at the outset can become very expensive in the long run.

Furthermore, repairing a plow during a wind-driven snowstorm with temperatures hovering around zero can be very uncomfortable. Be cognizant of such things as the size of the quick disconnects on the hydraulic hoses, the relative strength of these hoses, the size and location of the connecting pins and the ease of operation from the cab of the truck.

Plows now come with various quick attach and detach systems to allow for easy connection and disconnection of the plow from the truck. Unfortunately, with more moving parts comes more opportunities for things to go wrong or manfunction.

To Salt Or Not To Salt? Another service most snow professionals add as they grow is deicing, whether that be a simple salt spreader mounted on the back of a pickup truck or a sophisticated liquid deicing application system.

Most entry-level contractors purchase a salt spreader that can be mounted on the back of a pickup truck and then filled with bagged rock salt or another ice-melting product.

Employee Management. Overseeing all aspects of the business becomes impossible as the business grows. No matter how many hours you put in, you simply cannot do it all. This means that hiring and training employees becomes an integral part of your success and/or failure.

Finding good, qualified help is not easy if you offer low wages, skimpy benefits and less room for advancement than the "big boys." When starting out, you may not be able to offer such incentives; however, you'll need to strike a balance between what you can afford and what you cannot fail to provide. This is one big reason that employee relations is such an important part of small business management. Employees of small companies are often more satisfied than those working at large firms. Often they find their ideas are better received at small companies than large ones. They feel they are making a difference with the smaller company. The owner of a small business often respects his/her employees more, knowing "we are all in this together," so to speak.

However, responsibility and delegation of authority can be a touchy issue when dealing with small business owners (and often husband-wife teams). Long-term employees can often be difficult to deal with as the company grows and it becomes time to promote people to positions of authority. The small business owner feels loyalty to those who have been with him longest. Unfortunately, because someone has been with you a long time and is loyal to the cause is rarely a justifiable reason to heap authority on that person. Even though you may feel compelled to do this, you must (for the betterment of the company as a whole) step back and view the situation from a perspective of "how will this help the company overall?"

Most entry-level contractors purchase a pickup-mounted salt spreader for deicing.

The same can be said for family businesses that are expanding. The old saying that "you can't fire family" is just that – an old saying. Owners of small businesses can best serve themselves and their business if they gradually recruit and groom employees for management positions.

Eventually, you may have to allow your business to graduate

from an entrepreneurial company into a managerial one.

Documentation. One other thing that differentiates the small struggling business from the successful, thriving business is documentation. Too often the entrepreneur says he's too busy to keep track of everything. In the snow industry, record keeping is paramount to your success.

The most successful snow removal companies are very thorough about record keeping. They keep track of where people are working, when salt trucks arrive on site, how much material they use on the site, phone calls that come into the business requesting services and even outgoing phone calls to customers. Other information to be documented includes snowfall depths at various customer locations, temperature fluctuations during a snow or ice event, how many cars are in the customer's lot and special requests made by customers. Some contractors even go so far as to keep track of customer contact on site and contacts with anyone on a particular snow removal site.

As you grow and become more involved with insurance companies, adjusters and customers who complain about their invoices, you will realize the importance of record keeping. The more records you keep, and the accuracy of those records, can mean the difference between success and failure in any given snow season.

❄ THE PRICE OF SNOW REMOVAL

Getting started in the snow removal business can be as simple as a shovel or as sophisticated as a dump truck mounted with a box plow and liquid deicing unit. Here's a rundown of what a basic, entry-level snow business would need.

- Four-wheel-drive pickup truck:
  - New: $28,000
  - Used: $15,000
- Basic plow blade: .... $1,000 to $4,000
- Snowblower: .... $350 to $1,500
- Basic salt spreader: .... $1,000 to $2,000
- Bagged rock salt: .... $8 per bag

WORKING AS A SUBCONTRACTOR

Once you have the necessary equipment, there are several ways to enter the snow and ice removal industry. One common entry method is to become a subcontractor for an established professional in your area.

Working as a subcontractor is relatively easy and free of hassle. You generate income for yourself and you don't have any of the headaches associated with managing a business.

You don't have to sell yourself to customers. You don't have to price out the jobs. You don't have to worry about turning a profit from sales and service. You don't have to deal with customers. You don't have to collect money. As a subcontractor, you just do the work and get paid a fair amount for your efforts.

Working as a subcontractor also provides several of the benefits that come with self-employment. You can depreciate your equipment and deduct associated expenses. These legal deductions allow you to keep quite a lot of the money you earn.

Some contractors do quite well working for other snowplowing contractors in this fashion. A loyal subcontractor who loves plowing snow can generate a nice income working under the direction of a larger contractor who takes all the risks, finds all the work and handles all the documentation.

Finding A Pro To Sub For. Finding reputable contractors to work for is not as easy as it sounds. Unfortunately, there are a few unscrupulous contractors out there who will promise the world and then not live up to their own billing. There is usually at least one contractor in every market who operates this way. Ascertaining who these people are takes some effort.

When you seek out a business to subcontract for, research it as you would any potential employer (although you will not be an employee when you work as a subcontractor). The best sources of reference when seeking out someone to plow for are those who are already working for the plowing contractor. Usually you will hear who is hiring by word-of-mouth or through a newspaper advertisement in the "General Help" section of the local newspaper.

Remember, the interview process is a two-way street. The contractor will ask you questions, but you also can ask questions of your own. Some of the things you want to know are:

- How do you route your subs?
- What kind of customers does your company service (residential, commercial, retail, industrial)?
- How do you pay your subs?
- What kind of insurance do you require of your subs?
- Do you plow gravel lots?
- How much can I expect to work during any given snow event?
- Do you run a speed crew for minor events?
- How will I fit into the speed crews and/or clean-up crews you use after a storm is complete and most lots are cleared?
- What about breakdown time?
- What is your policy on travel time between jobs?
- Can I plow some of my own accounts too (not while you are paying me, of course, but I do want to grow my business as well)?
- What is your policy about who gets called out first?
- Will you let me talk to some of your other subs?

SNOW TERM: LAKE-EFFECT SNOW

During fall and early winter, as cold air flows over the relatively warm water of the Great Lakes, the air is heated and water from the lakes is drawn into the atmosphere where it forms clouds, becomes dense and falls as snow. The greater the temperature contrast between the air and the water, the more snow that falls.

Before subcontracting for a particular company, check the company's references to make sure no warning signs stick out.

A viable and honest contractor should have no problem providing you with references. The type of references you want are those who already plow for this contractor. Call a few of the references and chat about the policies of the contractor you are considering working for.

A good sign is when you can find many subs who have worked for the contrator for several years. Make sure the contractor has a generally good reputation among his peers. Further, make sure there are no unresolved issues existing with the local Attorney General's office or Better Business Bureau.

SNOW FACT: Based on 1961-1990 National Weather Service records, Rochester, N.Y., is the snowiest large city in the U.S. (population greater than 200,000), averaging 94 inches of snowfall annually. Buffalo is a close second.[2]

SNOW REMOVAL AS AN ADD-ON SERVICE

Rarely does a person begin his or her career as a business owner in the snow industry. More often than not, an individual contemplating the snow removal industry already owns a property-service-oriented business. The most common of these include landscape contractors, pavement maintenance contractors, excavation contractors, roofing contractors, pool installation contractors and other seasonal-type contractors, who enter the snow removal business to generate additional revenues in the off-season.

Sometimes these contractors enter snow removal because they need to keep existing customers happy or keep valued employees active during winter months, during the inevitable lull in work of the core business. Whatever the reasoning, snow and ice management should be viewed as a profit center and not as a necessary evil. If your core business already has some of the necessary equipment and staffing available, it may make sense to add snow removal to your business mix. However, some issues should be resolved within your business model prior to making this decision.

Staffing. One of the first considerations should be the staff of your existing business. Will they support such a decision? Better yet, can you get your staff to make the decision you want for you? One probable scenario to success would be to allow a higher-level management employee to research the possibilities prior to implementation of the snow business model.

If the support staff buys into the idea prior to you issuing a mandate, the entire process of adding snow removal will be easier to implement. The arguments against such an endeavor (from the support staff) usually result from ignorance regarding what is required to be in the snow removal business and regarding the attainable margins with proper planning and attitudes. As a result, make sure staff are properly informed.

The Power Of Networking. Networking with other successful snow contractors, through associations and industry trade shows and events, can give you insight about venturing into snow removal. Such involvement can also generate excitement for the process and alleviate fears associated with any new endeavor. As an add-on operation, snow removal and ice control operations will most likely generate higher gross margins than your core business, but only if approached properly.

SNOW REMOVAL AS A STAND-ALONE BUSINESS

While rare, some contractors are quite successful in operating businesses that do snow and ice removal as their only business. These contractors are typically located in areas with heavy snowfalls and they almost always have a lot of large commercial clients. This combination of intense winters and high-level clients makes it possible for such contractors to make a healthy profit, though only generating revenue for three to five months.

You can learn a lot by networking with other snow and ice management professionals.

For such businesses to work, however, the contractors must have deep relationships with reliable subcontractors, who own equipment geared at serving large-scale commercial/industrial clients. Otherwise the contractors have to own the equipment themselves, which makes earning a profit nearly impossible, especially because the equipment would sit idle during non-winter months.

These factors — reliance on subcontractors, need for long and heavy winters, and a roster of deep-pocketed commercial clients — make owning a stand-alone snow business a serious challenge and appropriate only for those with solid experience in snow removal and business management. ❄

SOURCES

[1]National Climatic Data Center, National Oceanic and Atmospheric Administration, National Snow & Ice Data Center, University of Colorado, Boulder, Colo., http://nsidc.org., AccuWeather.com, www.accuweather.com.

[2]*The Snow Booklet*, Colorado Climate Center, Colorado State University.

CHAPTER THREE: Becoming A Business

IN THIS CHAPTER

- Establishing yourself as a legal business entity
- The various forms your business can take

Assuming that you want to establish yourself as a business and operate independently, one of the first things that should be done is obtaining a federal employer identification number (EIN). An EIN is a nine-digit number the Internal Revenue Service (IRS) assigns to employers for tax filing and reporting purposes. You do not need to have employees in order to obtain an EIN number. EINs are free and they are easy to obtain. You may use your EIN on all tax returns, employment tax checks, employment tax returns and other employment documents that you send to the IRS.

You can obtain an EIN by filing IRS form SS-4 (Application for Employer Identification Number) with the IRS. Completing the form is simple and the form contains detailed instructions. You can obtain your EIN by mailing the completed form to the appropriate IRS service center listed in the form's instructions. The IRS will mail the EIN to you in about four weeks. If you need the EIN right away, you can get it via telephone by using the IRS's Tele-TIN program. The SS-4 form has instructions for this also.

Depending on the business structure, legally establishing your business can involve a little or a lot of paperwork.

Having an EIN establishes you as a business, and some companies you do business with will request this information. Your EIN number can be used on the standard W-9 form that you may have to fill out when dealing with companies that pay you for services rendered. You can use this EIN even if you are a sole proprietor. The EIN would be used in place of your Social Security number on the W-9. Some companies will look more favorably on your status as a business if you have an EIN.

Being self employed has many advantages, but there is also an overwhelming time commitment to being self-employed.

Customers can be quite demanding, and the work we do in the snow industry is often done at difficult hours under horrendous conditions, which takes time away from friends and family.

Anyone who tells you being self-employed is fun, you can take time off whenever you want and you have no worries – has never experienced a growing business that is "your baby." Also, if you get sick, get hurt or become incapacitated in any fashion, the business may suffer from your absence.

> **❄ SNOW TERM: BLIZZARD**
>
> 
>
> Blizzards are severe winter storms that pack a combination of blowing snow and wind resulting in very low visibilities. While heavy snowfalls and severe cold often accompany blizzards, they are not required. Sometimes strong winds pick up snow that has already fallen, creating a blizzard. Officially, the National Weather Service defines a blizzard as large amounts of falling or blowing snow with winds in excess of 35 mph and visibilities of less than one-quarter of a mile for an extended period of time (greater than 3 hours). When these conditions are expected, the National Weather Service may issue a "Blizzard Warning." When a less severe, but still dangerous, winter storm is expected a "Winter Storm Watch" or "Winter Storm Warning" may be issued.[1]

There are three basic business entities that are available to the snowplow contractor for his/her business. These are the sole proprietorship, partnership and the corporation. Each has its good and bad merits and we shall discuss each of these in turn.

THE SOLE PROPRIETORSHIP

A sole proprietorship is an organization that is owned and operated by one person. It is the most common form of business ownership in the United States. A sole proprietorship is easy to start and end. All you have to do to start a sole proprietorship is simply say, "I'm in business." The simplest form of announcement is satisfactory evidence that you are in business. The business is just as easily dissolved. You simply stop. There is no one to consult or to disagree with about such decisions (although your spouse and family usually have some input). There is pride in ownership and you do not have to share the profits with anyone else (except for mandatory taxes to Uncle Sam).

Interestingly, while sole proprietorships are the most common form of business ownership in the United States (72 percent of all businesses), they only represent 5 to 10 percent of the sales volume.

There are no special taxes associated with being a sole proprietor. All the profits are taxed as personal income; however, the owner does have to pay esti-

Though not without disadvantages, a partnership arrangement can be a good way to share the risk of running a business.

mated taxes quarterly. You cannot wait until April 15 to settle up with the government.

The down side is you cannot collect unemployment if times are slow, cannot deduct all of your health insurance premiums as business expenses, get no sick leave, no paid disability insurance and the like. For the paid worker/employee, these fringe benefits can add as much as 30 percent to their total wage compensation.

Proprietor Beware. There are other disadvantages to being a sole proprietor. When you work for another person, it's his risk that is problematic. As a sole proprietor, you assume all the risks of being in business yourself. Any debts you incur are your responsibility. This means if you mismanage the business you can lose your home, your savings, your personal assets and the like. This is a serious consequence of being self-employed and not one that should be taken lightly.

Should you die, the business ends unless your heirs take over or unless you have made provisions for the sale of the business after your death. Often, though, your customers are dealing with you (the sole ownership entity) and will not want to deal with anyone else should you die. This is a concern if you, your family and employees have begun to achieve wealth through operation of the business.

THE PARTNERSHIP

The partnership arrangement can have advantages over the sole proprietorship, but it too has its disadvantages. There are actually several types of partnership arrangements. These include general, limited and master limited partnerships. The general partnership is one in which all owners share in operating the business and in assuming liability for the business.

Pushing The Limit. A limited partnership is a partnership with

SNOW FACT: An average of 105 snow-producing storms enter the 48 contiguous states each year.[2]

NOTABLE SNOW-STORM: December 1778: In Rhode Island, drifts were reported to be 15 feet high after this storm.[3]

one or more general partners and one or more limited partners.

The general partner is an owner who has unlimited liability and is active in managing the firm. A limited partner risks an investment in the firm but enjoys limited liability and cannot legally help manage the company. By limited liability we mean that the limited partners are not responsible for the business's debts beyond the amount of their investment. Their liability is limited to the amount they put into the company. Their personal assets are not at risk.

Master Limited Partnership. The master limited partnership acts much like a corporation and may be traded on the stock market. However, a master limited partnership pays taxes like a partnership. In this way the business avoids the required corporate income tax. It is rare to see this type of entity in the snow and ice management industry.

Advantages. There are a few distinct advantages of the partnership arrangement over the sole proprietorship entity. For one, there is a pooling of resources that comes with having a partner. Some would argue it is easier to manage a business with a partner, a person to share information and management responsibilities with.

There are shared financial resources to be considered. The liability is shared, thus taking the burden off one individual to provide the financial wherewithal to finance growth and acquisition of equipment and inventory. Management of day-to-day activities is shared, thus the overall workload is distributed amongst the partners.

Disadvantages. The disadvantages of partnerships can be great. Anytime two or more people must agree on something, tension can ensue. Partnerships can destroy families, friendships and marriages. You must share the profits with others, even if one of the partners believes he is doing most of the work. Liability is shared, which means if a partner makes a costly mistake through ignorance or an error in judgement, all the partners will take the hit. General partners can lose their personal assets and not be the one at fault.

Disagreements amongst the partners can be cause for dissolution of the entity. And, if such dissolution of the entity is to take place, there can be considerable argument over who gets what. It is best to decide prior to going into business together how the entity will be split up in the event of dissolution. It's almost like having a prenuptial prior to getting married, which in fact, a partnership is very much like. Do not jump into a partnership relationship without considerable due diligence.

THE CORPORATION

While the word corporation will make one immediately think of Microsoft or IBM, small businesses often opt for corporate status as a business entity. A "C" corporation is a state-chartered legal entity with authority to act and have liability separate from its owners. This means that the stockholders (owners) of the corporation are not responsible (liable) for the debts or any problems of the corporation beyond the money they invest in the corporation. As a result, the stockholders no longer have to worry about losing their homes and personal assets.

A corporation not only limits the owners' (stockholders) liability, it allows many people to share in the ownership (and profits) of a business without working there or having any other management commitments to the business.

In most cases of snowplowing companies, the stockholders are the owners and often do the majority of the work necessary to keep the business operating. This limited liability is the major reason for establishing a corporate status.

Additionally, the corporation becomes a perpetual entity that can survive the death of any of the stockholders. It is also easy to change ownership of the entity as a stockholder simply sells his/her stock to someone else. And, it is possible to separate ownership from the management of the corporation.

Disadvantages. There are some disadvantages to the corporate setup. The initial cost can be expensive if you involve lawyers and accountants; however, there are now easier and less expensive ways to do this yourself in some states. There is a lot of paperwork associated with corporation structures.

A corporation protects the owners from certain financial liability, but involves increased record keeping.

Sole proprietorships can keep less stringent records, but a corporation has to abide by tougher stringent recordkeeping rules. These rules include filing two tax returns. The individual who owns the stock in the corporation must file an individual tax return; however, the corporation must also file a tax return (and this can be quite complex if the corporate entity is large). This often results in double taxation for the individual.

Further, states often tax corporations more harshly than other entities discussed herein.

Keep in mind that a corporation does not need to have many

owners and does not need to have hundreds of stockholders. Individuals may incorporate themselves to take advantage of the liability protection that is afforded individuals who hold stock in such entities. Many firms incorporate in Delaware simply because it is easy to do so.

How to Form Your Own Corporation without a Lawyer for Under $500, by Ted S. Nicholas (published by Enterprise Publications) is an excellent source to find out the steps you need to follow to establish your business as a full corporation.

S CORPORATION

An "S" corporation is a unique government creation that looks like a corporation but is taxed like a sole proprietorship. S corporations have the benefit of limited liability and the "paperwork" is much like a regular corporation. S corporations have shareholders, directors, employees, etc., but the profits are taxed as the personal income of the

❄ SMALL BUSINESS RESOURCES

Managing the business side of a snow and ice removal company can be very hard work. Luckily for snow professionals, there are many Web sites with helpful information. Below are listed some of the best Web sites for snow business owners:

- www.sima.org
- www.sba.gov
- www.fsb.com
- www.sbsc.org
- www.sbnpub.com
- www.business.gov
- www.business.com
- www.smallbizmanager.com
- www.ceoexpress.com
- www.tsbj.com
- www.entrepreneur.com
- www.irs.gov/smallbiz
- http://smallbusiness.yahoo.com
- www.quicken.com/small_business
- www.businessweek.com/smallbiz
- www.lawnandlandscape.com

shareholder(s). This avoids the double taxation of regular corporations; however, S corporations must pay self-employment taxes in the same manner as sole proprietorships.

Not every business can become an S corporation. To qualify a company must:

- Have shareholders who are individuals or estates and are citizens or permanent residents of the U.S.
- Have only one class of outstanding stock
- Not own 80 percent or more of the stock of an other corporation
- Not have more than 25 percent of income derived from passive sources (rents, royalties, interest, etc.).

LIMITED LIABILITY COMPANY

A recent addition to the business entity family is the limited liability company (LLC). This is similar to the S corporation without the special eligibility requirements. The sole-member limited

liability company provides liability protection for sole proprietorships. LLCs have all the flexibility of partnerships with the same kind of liability protection provided regular corporations. With an LLC, you cannot sell stock, but profit is taxed as personal income. Limited liability companies are becoming very popular for people in the snow industry.

As you can see there are a number of options for your business. Work with your accountant and attorney to choose the one that fits best with your situation and goals in business. ❄

SOURCES/FURTHER READING

[1]National Snow & Ice Data Center, University of Colorado, Boulder, Colo., http://nsidc.org.

[2]The Snow Booklet, Colorado Climate Center, Colorado State University.

[3]National Oceanic and Atmospheric Administration, National Snow & Ice Data Center, University of Colorado, Boulder, Colo., http://nsidc.org. AccuWeather.com, www.accuweather.com, National Climatic Data Center.

How to Form Your Own Corporation Without a Lawyer For Under $500, by Ted S. Nicholas, Enterprise Publications.

CHAPTER FOUR:
Finding Customers

IN THIS CHAPTER

- Securing customers for new business
- Marketing techniques for established companies
- Developing business relationships
- Promotion and advertising

No business exists without customers. So for any snow professional, new or experienced, finding customers should be a chief focus.

ADVICE FOR STARTUPS

Once the decision has been made to enter the snow removal industry and after acquiring the necessary equipment, one needs to secure customers to generate revenue. You have to find customers and get them to notice you. Straightforward approaches usually work best.

Canvassing Neighborhoods. One method would be to target a neighborhood and distribute flyers that promote your company's winter services. Placing several hundred flyers in a neighborhood that appears to be affluent enough to warrant snow and ice removal services is an ideal place to start. Include a short description of the services offered, your name (or the name of the business entity) and a phone number to call — inevitably you will get some calls.

The timing is important. Potential residential customers are probably not interested in thinking about snowplowing in July. Waiting until the temperatures drop can increase the interest in your services, as the declining temperatures are likely to bring winter into the minds of your potential customer. Be ready to "get the word out" the day after the first really cold night. Remember that you cannot put such flyers in mailboxes as this is strictly forbidden by the U.S. Postal Service and can get you into some considerable difficulty with federal authorities.

Brochures can be a relatively low-cost sales and marketing tool.

The Customer Meeting. Once calls start to come in, make appointments with the potential customers during which you can

The face-to-face customer meeting is a key to gaining business and earning the customer's trust.

pitch your company. Have business cards to distribute upon meeting these potential customers. Business cards are very inexpensive, and you can even print them yourself with the software programs that are readily available at local office supply stores. Business cards are a fabulous way to get your point across.

A basic must have — business cards.

Always be honest with potential customers. When speaking to the prospective customer, explain that you are just starting out. If there is any resistance, accentuate the positives of being small, which is really quite easy. And you should practice the "pitch" in the shower or in front of a mirror.

Beware the potential customer who tells you, "If you give me a good price, I'll get you a lot of work in this neighborhood." Rarely do these promises come to fruition. Being fair and honest will get you more new business than low prices. Always price your jobs to make a fair and honest profit.

MARKETING FOR ESTABLISHED COMPANIES

Finding customers for your snow and ice removal operations is decidedly less difficult if you have an existing customer base, say from an existing nonwinter company like a lawn maintenance firm. A first step is to contact all of your existing customers. Studies have proven that a high percentage of any good service business' new revenues come from existing and past (satisfied) customers.

A simple flier announcing the new snow and ice service you are offering is easy to generate and easy to get into the hands of existing and past customers. A mailing to past customers (assuming you keep records of who you have dealt with) will generate interest.

If you have a sales staff, they should follow up by phone to inquire about the possibility of servicing these past customers in this new endeavor. If these people are already under contract for snow removal and if they are a satisfied past customer, ask questions. Have a form ready for such calls so you can gather information that will be helpful to you in future sales calls.

Past customers are often easy to gain information from in this fashion. Ask if they are satisfied with their current service providers. If they are, press further. Tell them it is good they have good providers, but ask if they would help you learn how best to provide good service to other customers? For example, say, "That's great, I'm glad you are satisfied. I won't try to pressure you into switching providers. However, we want to provide the same high quality service you are currently accustomed to. Can you help me by educating me on what you consider a good service provider?"

Assuming an affirmative response, continue to ask questions:

- What level of service do you demand?
- How much are you paying for snow and ice removal?
- What kind of plowing equipment does your current provider use?
- Who is this good contractor?
- What are the contractor's billing procedures?
- What makes them a good contractor?

Mark down the answers. They are important for several reasons. You are developing a picture of what this customer perceives a good contractor to be and you are finding out pricing structures. In essence, you are doing a marketing survey.

Survey Says. Another avenue to gain information is to send a questionnaire to existing customers asking them if snow removal as an add-on service would be of value to them. If this is presented as an information-gathering endeavor, and not a solicitation for business, you may be pleasantly surprised by the response. People love to be asked their opinions. They often want to help and are often easier to approach than most of us think.

Networking. Along these lines, it would be wise to seek out the

❄ DON'T SWEAT THAT I'M SMALL

Small and/or new snow removal companies can use their size as an asset to compete against larger companies. Here are a few pitches small companies can use to market to customers:

- "Yes . . . I'm just starting out. But that's really a benefit to you. I really want to please my customers and I'll pay very close attention to your needs."
- "Yes . . . I'm new in the business, but that's really a benefit to you. I don't have hundreds of customers, so I'll be very attentive to your needs."
- "Yes . . . I'm new to this market, but that's really a benefit for you. I don't have to cater to lots of customers, so you'll get the attention you deserve."
- "Yes . . . I'm new in the business, but that's a benefit to you. I'm honest, fair, hardworking and I want to please you so you'll tell your friends to do business with me."

best snow contractor in your market and talk to him. People love to talk about themselves. If you approach a reputable snow contractor, the chances are very good he will be much more open

Consider asking larger snow contractors in your market for referrals.

to you if you tell him up front that you are considering entering the business and would like his advice. No matter if he likes or dislikes the business, they have valuable information you can use.

Be honest in that you will not be able to compete with him at his level, but if it should happen, you would like to be an honest competitor and not be branded a low-baller.

In seeking the advice of an experienced person in the industry, you automatically set yourself apart from the "plow jockey."

Most businessmen would much rather educate the person just starting out, as it will keep you from degrading the industry with poor business practices.

You should also consider asking larger contractors in your market for referrals. All large snow removal contractors stop taking on new business at some point, as it is good business not to overbook. It's also good business to refer inquiries to another viable contractor. Good businessmen realize that to just turn away a potential customer is bad business strategy.

Approach larger contractors with this in mind, and you may end up with a tremendous book of business in a very short time. Some larger contractors can make a small contractor viable just through referrals. Although if this happens, don't forget who helped you when you were small. Favors should be repaid at some point in the future.

There are other practical avenues to secure new customers and referrals. Hand out business cards to everyone you meet. Asking for referrals takes little effort and is often overlooked during the service process. Most people won't ask for a referral simply because they have thin skin. Remember, the word "no" will not hurt you physically.

And don't forget that old standby — the cold call. Soliciting business is not bad business. It does take some time, and you will get turned away, but asking for business is an age-old sales tool that works. We will deal with some specific sales attitudes in Chapter 5.

OTHER MARKETING METHODS & MATERIALS

You must establish your company's own identity and determine how you will service your customers. Once determined, communicating this message to potential customers is defined as "marketing."

Marketing is many things to many people. Most small businesses deal with marketing as a means to an end — the end result is to secure new business and retain it for the long term. Marketing is not sales, nor is it a substitute for a good sales presentation. Marketing is defined as the process of determining customer wants and needs and then providing these customers with goods and services that meet or exceed their expectations.

Defining the wants and needs should be fairly easy, however, providing the customer with the service that will meet those expectations takes a bit more work. The challenge is to define what any particular customer's expectations might be. Some, as in the industrial sector, may not require bare pavement (pavement completely cleared of snow), as they only need to get employees into the manufacturing building and don't have constant traffic.

Retail customers have the greatest liability exposure and the most traffic into their facilities, thus requiring a higher level of service. Some customers require 24-hour-a-day, 7-day-a-week service. This requires a tremendous commitment on the part of the snow removal contractor, as equipment may need to be manned all the time in the event of a catastrophic snow event. Your own personal commitment level may decide what level of service you can, or will, supply to your customers.

Decisions, Decisions, Decisions. With the above in mind, some decisions need to be made. What quality level of service do you desire to achieve and provide to potential customers? Having your name (or the company's name) known for professional quality service is much more desirable than being known for cheap prices. Nothing of value comes cheap and having an identity that denotes cheap should not be anyone's goal in this industry. If you can lower a customer's expenditures by achieving better productivity, this might be a viable reason for potential customers to deal with your company. However, one must have an identity in order to have a reputation.

This is where marketing comes into play. Marketing your company's snow removal services is how to help in establishing an

NOTABLE SNOWSTORM: November 1798: "The Long Storm" went down in history as the snowiest on record for November. Stretching from Maryland to Maine, up to a foot-and-a-half of snow coated the region.[1]

Effective communication and follow-up is crucial to maintaining long-term customer relationships.

identity. If the goal is to have a reputation and identity that you provide quality service and protect your customers liability exposure (and their customers' in the case of retail establishments), then you need to market your company along those lines.

The most important method is to be able to back up what you tell people. Establishing a reputation for having bare pavement properties when you have completed your work takes some time. However, you can assist that effort through effective marketing of those values. Let's assume that you wish to establish a long-term relationship with any good customer.

SNOW FACT: Buffalo, N.Y. is a close runner-up in terms of U.S. large cities with the most snow. A 39-inch snowfall in 24 hours in early December 1995 cost the city nearly $5 million for snow removal.[2]

SUCCEEDING WITH RELATIONSHIP MARKETING

Relationship marketing is defined as establishing and maintaining mutually beneficial exchange relationships with customers (internal employees and external paying customers) and all others involved in and with your organization. Relationship marketing moves away from the traditional mass marketing concepts toward custom relationships designed to satisfy specific areas of the overall market.

Snow removal and ice management services fall under this definition because most of our potential customers have specific wants and needs to satisfy. In this fashion you can work at retaining customers instead of constantly finding new ones. It should be easy to see that retaining customers is much more favorable than constantly turning over customers in hopes of finding one cash cow that will make all your problems go away. That simply doesn't happen.

Some professionals believe that retaining 2 percent more customers has the same effect on profit as cutting costs by 10 percent. Relationship marketing means establishing and maintaining long-term relationships with customers.

One of the most valuable lessons you can learn is that you cannot take a shotgun approach to the marketing of your company. You cannot be *all things to all people*.

So, decide if you wish to concentrate in the residential snow markets or the commercial snow markets. We shall discuss the merits of both later in this book, but it is a decision that must be made at some point.

PROMOTION

The established promotional tools available to a business entity are:

- Advertising
- Word of mouth
- Sales promotion
- Public relations
- Publicity

Many people equate promotion with advertising because they are not aware of the differences between the promotional tools noted above. Advertising is paid, non-personal communication through various media by companies and individuals who are in some way identified in the advertising message. Word of mouth is not a form of advertising because it doesn't use an alternative medium (newspaper, Yellow Pages, radio, television, etc.). It's not paid for and it's personal. Publicity is different from advertising in that media space for publicity is not paid for. Personal selling is face-to-face type communication and doesn't go through another medium and thus is not advertising.

Advertising. Advertising is a method of "getting the word out." Yellow Page advertising is a common form of letting the buying public know you offer snowplowing services. However, there is some argument that this may not be the best form of advertising for snowplowing, as it invites everyone who gets a telephone book to call you, no matter how close or how far they may be from your office. Due to the nature of this business, getting inquiries from many miles away from your base of operations often results in having to turn away the business. Snowplowing by definition is regionalized and difficult, but not impossible, to operate over a large geographical area.

Advertising in the newspaper can narrow the focus of your efforts. Television advertising usually widens the scope, however this can be inefficient from a cost analysis standpoint. Television advertising is usually expensive and may not offer an adequate return on your investment.

Internet advertising is developing rapidly, but again, due to the large potential audience for such advertising, this medium may not yet be beneficial due to the localized nature of most snowplowing business.

Yellow Page advertising is commonly used, but is not narrowly targeted.

Spread The Word. Word-of-mouth promotion is usually the most preferred method of promoting the business. This form of marketing is free, has inherent credibility and requires very

---

❄

SNOW TERM: SNOW BURST

Very intense showers of snow, often of short duration, that greatly restrict visibility and produce periods of rapid accumulation.[3]

SHOESTRING SNOW SELLING

Low cost sales and marketing strategies for snow and ice contractors.

Successful marketing doesn't have to mean high-budget television and radio advertising spots or glossy brochures and direct mail pieces. Many savvy contractors say it's not *how much* money you spend, but *how* you spend it that counts.

Take Erv Denig, owner, Lawn & Turf Landscaping, Fort Wayne, Ind. Denig grows by one crew a year, he said, through highly targeted, face-to-face selling. "We don't advertise, we don't do direct mail, we go after the client we want and the client that fits our territory," he said.

Other than a small Yellow Page advertisement, the company relies solely on face-to-face sales meetings with potential clients. Each year, Denig identifies new commercial facilities he thinks would fit his snow operation and evaluates the potential job site. Denig makes an informal inventory of the facility by documenting the potential snow and ice removal needs and photographing damage that a previous contractor may have caused.

Denig asks office personnel at the account about who makes the decisions regarding grounds maintenance, how long the decision maker has been with the firm and which contractor conducted winter services for the facility last year — all this before ever formally calling the prospect. When he's gathered all the facts, Denig calls the decision maker, briefly discusses the facility and its snow removal needs and requests a face-to-face meeting. By completing his pre-call research before approaching a client, Denig said he's able to set up face-to-face meetings with nearly 100 percent of prospects.

During the sales meeting, Denig promotes his company's professionalism through customer testimonials and a list, including pictures, of the company's snow removal equipment. "When they see our lineup of equipment, that usually convinces clients they should have us, rather than some pickup truck operator," Denig said.

WHAT MAKES YOU DIFFERENT? Marketing and sales consultant Fred Berns, president, Power Promotion Inc., Lafayette, Colo., says a key to maximizing your marketing dollar is determining the unique attribute that sets your company apart from the competition. "Become the number one expert on you and your company. Ask yourself, 'What makes me outstanding at what I do?'

"Everyone you are trying to influence is subjected to approximately 1,250 sales messages a day. So why should they listen to your message? It's a real important question."

Berns said contractors can discover their differentiating attribute by thoughtfully considering a few of the following questions. What do I specialize in? What are my special skills? What are my accomplishments? What kinds of clients do I specialize in? What special degrees or certifications do I have?

"I think you can set and get any price you want, as long as you can sufficiently differentiate yourself from competitors. Really focus on what makes you different," Berns suggested.

Once you determine the items that distinguish your snow

removal operation, there is a virtually limitless supply of relatively low-cost ways to promote this message to your market. Ideas include distributing snow and ice tips to potential customers, sharing tips at community and local organization events, sending stories and press releases to local newspapers and including your company's unique selling point in your voice-mail message. "Americans spend approximately a total of 3.2 billion hours a year talking and listening to voice mail. If your voice mail is not special, you are missing out on a major marketing opportunity," Berns said.

Other ideas include creating a customer satisfaction survey, hosting focus group meetings with small teams of customers or hosting a customer appreciation day.

Such customer events can be a prime opportunity to seek the advice of clients on your service. Such events also present a great opportunity for referrals and testimonials.

FIVE-MINUTE MARKETERS

Consider these time- and cost-cutting marketing ideas:

- Update your voice-mail recording to include a promotional message that communicates your unique services.
- Ask a client for a referral and a testimonial letter.
- Ask a client for additional business.
- Contact a former client to renew the business relationship.
- Determine three ways that you differ from your competitors.
- Meet five people at a networking event in your market.
- Distribute 10 business cards at a local trade show.
- Get on the mailing list of top prospects and competitors.
- Invite a decision maker to lunch.

Source: Sell Yourself! 501 Ways to Get Them to Buy From You by Fred Berns. For information visit the Web site: www.fredberns.com.

FOLLOW-UP. Steve Bednarz, vice president, Tovar's Snowplowing, Elgin, Ill., says the simple act of following up with a prospect can be a powerful, low-cost strategy to earning new business. After a direct mail piece is sent, Bednarz contacts potential clients to make sure they've received the brochure. Bednarz also makes sure he follows up via phone after a face-to-face meeting and, most importantly, after a bid has been submitted.

"We try to have a good system of follow ups, so we don't just drop off a bid and never talk to the client again."

Follow up, Bednarz said, allows the snow contractor to uncover the prospect's thoughts on the proposal and make additional strides to win the business. "We can find out what we need to do to make it happen," Bednarz added.

Whether it's asking customers for referrals or following up with prospects after the initial bid, there are hundreds of low-cost marketing strategies available to contractors. The key is making the effort to implement them. ❄

Reprinted with permission of Snow Business *magazine.*

little time investment. By its very nature, word-of-mouth advertising involves people telling other people about the products and service providers they favor. When an individual recommends your company, the referral becomes an unpaid testimonial of your services. People often trust their neighbor's advice more than your marketing.

To achieve good word-of-mouth promotion, a company needs to focus on customer satisfaction. The company must deliver on promises made in the sales process, and focus on those customers who will, in fact, talk about the quality of your work.

Unfortunately, most of us do not use this method of self promotion effectively.

Publicity/Company Image. A presence in the community or your niche market through name recognition can be an important component to keeping and finding new customers. Publicity can involve something as simple as putting tasteful, yet promotional graphics on your vehicles. Promotion can also involve sponsoring a community function or event, such as a baseball league (see photo, above left). Identify your company's niche and then use those opportunities where you will be seen by your niche audience. For example, if you target commercial customers, exhibit at a chamber of commerce tradeshow or event. ❄

SOURCES/FURTHER READING

[1]National Climatic Data Center, National Oceanic and Atmospheric Administration, National Snow & Ice Data Center, University of Colorado, Boulder, Colo., http://nsidc.org. AccuWeather.com, www.accuweather.com.

[2]National Snow & Ice Data Center, University of Colorado.

[3]National Snow & Ice Data Center, University of Colorado.

The World's Best Known Marketing Secret: Building Your Business with Word-of-Mouth Marketing, by Ivan R. Misner.

CHAPTER FIVE: Sales

IN THIS CHAPTER

- ❄ Selling to residential customers
- ❄ Selling to commercial customers
- ❄ Sales strategies
- ❄ Bid vs. negotiate

One of the major questions that arises when plowing contractors get together is, "How do I get new customers, with all the cut-throat, one-truck operators out there giving away their time?"

Maybe this is the wrong attitude. Might it be better if we (as the contractors providing services) were in such a position that we interviewed customers who we wanted to have instead of quoting work hit or miss? Targeted selling is a key to success in any business. Why should it be any different with the snowplowing business?

SELLING TO RESIDENTIAL CUSTOMERS

If your focus is to grow your business by gaining residential customers, then you have to target these customers in such a fashion that will entice them to do business with you. Accentuating your commitment to perform quality work at a fair price will go a long way with residences that watch every dollar very closely. While residential customers are generally easier to obtain, they may be harder to keep. These customers can be fickle and change vendors quickly. It takes a special attitude to properly service residential customers.

A special attitude is required to successfully work with residential customers.

Residential customers want to know they are special and that you are going to cater to their

Apartment and condominium developments have traits of both commercial and residential accounts.

needs. Thus, marketing materials need to be geared toward that end. And, the residential market can be quite profitable if proper scheduling is kept in the forefront of the contractor's mind. Bundling such accounts in order to keep travel time to a minimum is the key to success in the residential plowing business. If you can plow a driveway in six to eight minutes (including travel between driveways) it is possible to plow up to eight customers an hour. If you wish to generate $200 per hour per vehicle, you can charge the customers $25 per push and realize some terrific margins for the company.

Word Of Mouth. Using existing residential customers as your springboard to more business is an intelligent method of using word-of-mouth advertising to one's advantage. Ask existing customers for referrals to their neighbors. Testimonials from neighbors work extremely well when soliciting new business. And, pricing structures that are fair and honest, as well as competitive, will assist in giving credibility to your cause.

An effective direct mail campaign to target neighborhoods can generate inquiries that can lead to sales. Such direct mail campaigns mean not just one mailing to each target residence. It takes at least three such contacts in order to have your name recognized by the customer, so you must be diligent in your endeavor. Include a postcard that asks the potential customer to reply to your office in order to secure pricing for your winter services. Once you get that inquiry, your sales ability should take over.

SELLING TO COMMERCIAL CUSTOMERS

For commercial customers, the same positive attitude should

prevail. If you have a shopping plaza you are servicing during winter months, then it would be in your best interest to have additional business close to that location. Assuming you are providing tremendous service to your own customer, then pay attention to what is happening across the street.

If that lot is not plowed on time, make a mental note of this. In spring, send a letter to the person in charge of that other site, requesting an opportunity to speak with him about his plowing needs. Do this in the spring so that the problems of the past winter are still fresh in his mind.

You may not get an opportunity to secure that business in the spring, but by staying in touch with the potential customer you should be able to arrange a meeting when the time comes to renew plowing contracts.

You might consider targeting all potential customers in the immediate vicinity of your satisfied customer. Ask your existing customer for a letter of recommendation that you can use when selling to customers in the same neighborhood or commercial development area.

If you are charging on a per-push basis, having new customers in the immediate vicinity will cut down on travel time between jobs. If you are charging per hour per truck with a minimum travel time charge between customers, it will mean increased revenue per truck.

Additionally, it is easier to incorporate a new customer into the schedule if it is close to an existing site you currently service. It just makes good sense to cluster your accounts strategically. Avoid taking on customers who have gravel parking lots

SNOW FACT: Snow kills hundreds of people each year in the United States, from traffic accidents, overexertion, exposure and avalanches.[1]

Commercial customers are lucrative, but demanding.

unless you can charge for the increased time it takes to plow such lots. Or, you might have to include a disclaimer that allows you to charge the customer to redistribute the gravel in the lots at the beginning of the spring. Another method of dealing with gravel lots is to use polyurethane cutting edges on the plowing equipment. Polyurethane edges virtually eliminate the disturbance of gravel in the lot, provided the lot has had time to freeze prior to plowing operations being started. However, extreme caution should be exercised when considering taking on such accounts.

NOTABLE SNOWSTORM: January 1857: "The Cold Storm" produced severe blizzard conditions along much of the eastern seaboard. Temperatures fell to minus 9 degrees F, and snowfalls were between 1 and 2 feet deep.[2]

SALES STRATEGIES

When a potential customer calls to get a price for plowing services, some contractors want to know why that customer is considering changing vendors. It is a fair question. If the customer is unsatisfied with the service provided by the previous contractor, this is a customer worth spending time with in order to ascertain the reasons for being unsatisfied.

You might find that the previous contractor was undercharging for his services, thus necessitating the need to short-cut the job in order to make a viable profit. In these cases you may need to be frank with the customer and tell him that he was not paying enough for the service and you are going to be more expensive, but the quality of work will be better. A potential customer who wants you to provide better service at the same price is not looking for quality and dependability. He is looking for the cheap price.

If he is just looking to check pricing, then you may want to avoid getting involved with such tactics, unless you need the practice quoting work without getting anything in return. Price shoppers will change vendors next year or, worse yet, mid-season just to get a cheaper price. In this industry, the cheap price usually ends up being un-justified by the second significant snowfall when the service just isn't there.

Focus On Your "Good Side." When talking with potential customers, you should be quick to share your unique selling point (USP), the quality that makes your company different, and better, than the competitors. If you are a large contractor with a large fleet of equipment, you might point out there is no excuse for not showing up on time (if enough snow falls soon enough to allow you to complete the plowing in the allotted time frame).

You probably have a mechanic on staff (or readily available) to get breakdowns fixed quickly so that the equipment is back up and run-

Your biggest sales point is professionalism. You provide a necessary service with specialized equipment.

ning in short order. You may have a full-time dispatch team to ensure special requests can be addressed in a timely fashion. All these points should be shared with the customer.

For small contractors, sell the fact that you don't have a large number of customers – so that every customer gets "personalized service." You don't have to keep track of a large contingent of trucks, so you always know where everybody is working. Personalized service means you care about that customer's needs, almost exclusively.

Remember: You Are A Professional. Don't hide the fact that you are making a profit at providing this service. While it may seem like everyone is plowing snow, the fact is you are in the minority. You provide a necessary service that requires specialized equipment, special talents and tremendous dedication to your customers. They should be glad you are there – ready and willing to provide services under terrible, and often unsafe, conditions. Sell these benefits of dealing with your company, no matter what size fleet you run.

Most importantly, keep in mind that you are in the snow and ice management business year round. While most of our customers only think of snow just before the season, we should be thinking about snow and ice all year long.

When you are asking about the landscape maintenance business, ask about the snow business too. When you do a landscape installation (or irrigation installation, paver installation, etc.) ask who does the snow re-

THE SALES PROCESS

Sales efforts directly impact the growth of your business through the procurement of new customers. There are several distinct steps.

1: PROSPECTING

Research potential customers and choose those most likely to purchase the snow services your company is offering. These are your prospects. These can include (depending upon the focus of your efforts), residences in a particular neighborhood; commercial and/or industrial accounts; retail establishments; hospitals and health care facilities; and condominium associations. You can glean the names of such customers as easily as riding down the street and paying attention to the establishments you drive by. Or, you might survey your own existing established customers to determine who is a viable prospect.

Don't forget that every customer you currently service is very valuable for your new venture.

Before actually contacting potential customers, or before making an on-site sales call, you should research prospects. Learn as much as possible about potential customers:

- What do they do?
- What are their snow removal needs?
- Which contractor are they using?

Such research is probably the most important part of the entire sales process. Walking in blind can prove frustrating if you are attempting to satisfy a need that is not really there. In other words, do your homework.

2: APPROACH THE CUSTOMER

Remember you don't have a second chance to make a good impression. If the first contact is to be in person, appropriate dress and a simple but explanatory business card is important. The idea is to impress the potential client with your professionalism, credibility and ability to satisfy the client's snow removal needs.

3: SALES PRESENTATION

After initially approaching the customer, you should formally present information about your company and what it can do to satisfy this customer's needs. Here you will need to show the advantages of doing business with your company as compared to the competition. In these instances, refrain from using the words

"I" and "me." An example of a poorly worded statement would be "I want your business." No kidding. That's why you are there. Customers need a reason to change how they currently do business. You'll need to ferret out why this customer should change vendors. Remember, the pain that comes with changing vendors needs to be considerably less than the pain of staying with the current vendor.

At this time, you'll need to determine if this customer has any reasons to make a change to your company. If all he is looking for is a way to lower his cost, thus making you lower your price below that which he is currently paying, this may not be a valid reason for you to take on the business.

4: OBJECTIONS

Your job is to make sure the customer has a valid reason to change vendors. Here you may experience objections from the customer. Don't view this as a method for the prospective customer to shoo you out the door. This is where the customer is attempting to quantify what you can do for him. If the customer is presenting objections to you (usually in the form of questions), he is talking to you. Answering his questions, and objections, is a give- and-take discussion forum. This is good. Avoid frustration and answer his questions in an honest and forthright manner. Do not ever lie to a customer. An experienced buyer will pick up on this and not trust you. Remember, if the truth can kill the deal you never had a deal.

5: CLOSING THE SALE

The next step in the sales process is "closing the deal." Here is where you must ask for the sale. Seven out of 10 presentations end without the sales person ever asking for the business. Often, that is the difference between securing a new customer and giving the business to your competitor. The biggest reason someone doesn't ask the question is that he is afraid of the word "no." And, if that word comes up, the best response is "why?" At that point you are back to the "objections" phase of the process. This is good. He is talking with you again. Ask "why" and then be quiet. Let the customer talk.

6: DELIVER THE GOODS

The sales process isn't complete until the service is delivered. This probably won't happen until it snows. Then the real test begins. Make sure you can deliver. Interestingly enough, this is then the first step in establishing a relationship with this customer. Hopefully it will be a lasting and rewarding relationship. ❄

moval. Is the customer happy with the service he is receiving? If he says yes, then that is good for your customer. Tell him he is lucky to have a good contractor, but if he becomes unsatisfied, then could you please have another opportunity to secure the snow removal business?

Establishing a quality reputation can go a long way toward securing new business.

Also, if he is happy with their current contractor, ask what he is being charged. If for no other reason than to see what the competition is charging, you should always ask. It's no threat to the incumbent, as you know this customer is satisfied.

Write down this number somewhere where it will not get lost. In a couple years, if you get a call to quote the work, you will have some idea "where the numbers are at" for that particular customer.

BIDDING VS. NEGOTIATING

It is the experience of the author that large customers who know your reputation for quality service and fair pricing will tend to lean toward negotiating pricing structures instead of bidding out the work. As a result, establishing a reputation for quality work and customer service can assist you greatly in your endeavors to secure new business.

Approaching the pricing question with a large customer can be tricky. Of course, he will want to be sure the pricing is competitive; however, the final pricing structures are often a result of negotiating a fair and honest deal whereby both parties win.

The agreement must be a "win-win-win" situation. The customer has to win, the people doing the work have to win and your company has to win. If all parties are not winners, then someone loses. Unfortunately, it seems that if one party loses, all parties lose.

Don't Be Outsold. The difference between good plowing contractors and great plowing contractors is their approach to sales and negotiation. Great contractors find a way to negotiate the contract with the large, potential customer instead of simply bidding the package and hoping.

The author of this book has a philosophy about sales that has allowed his own company to ascend to quite a high level in the industry.

This philosophy states that there are only two reasons to lose the sale: 1) either you cannot do what the customer wants done; or 2) you have been outsold.

This is a simple philosophy, and if you think about it, I think you will find that the philosophy is very true.

Price is never the overriding reason for not selling a customer on using your company. The price must be competitive, but there is no real value in cheap pricing. Most customers understand this, even if it is a subliminal idea that rarely comes to the surface. If price was the overriding factor in selecting a product or service – everybody would be driving Yugos.

CUSTOMER RETENTION

Numerous studies have shown that keeping an existing customer is much cheaper than finding new customers. With that in mind, some consideration needs to be given to this fact and what can be done to retain good customers. Even though you have provided great service at an affordable pricing structure, often a simple thank you will suffice to keep customers happy. A thank you card may seem basic, however, a well-written note can go a long way to satisfying the customer's need to feel wanted and needed.

Personalized thoughts on paper show the customer you really do care about his business. Additionally, a personal visit in the summer to say thank you will do wonders for the customer's attitude and perception of you.

One very successful method of retaining customers is through the use of multi-year contracts. Three-year contracts are normally sufficient. Customers do not have to go through the process of securing pricing each season, and you can better plan for your growth knowing you have an existing customer base to work from each season.

Some customers will agree to a 5-year contract, which further solidifies the bond between customer and vendor. This will also save considerable time when it comes to writing renewal contracts. This means you will only have to contact one-third of your customers each year to renew, leaving more time to solicit new customers and plan for your growth. Can customers break a three-year commitment? Yes. However, if you are providing good service and communication, customers will stay with you from year to year. ❄

SNOW TERM: WINTER STORM WATCH

Alerts the public to the possibility of a blizzard, heavy snow, freezing rain or heavy sleet. Winter Storm Watches are usually issued 12 to 36 hours before the beginning of a winter storm.[3]

SOURCES

[1]National Climatic Data Center, National Oceanic and Atmospheric Administration. National Snow & Ice Data Center, University of Colorado, Boulder, Colo., http://nsidc.org. AccuWeather.com, www.accuweather.com.

[2]National Snow & Ice Data Center, University of Colorado.

[3]U.S. National Weather Service.

CHAPTER SIX: Insurance Coverage

IN THIS CHAPTER

- ❄ Avoiding risk
- ❄ Types of insurance

Insurance coverage for snowplowing contractors is a must and not having insurance to cover potential losses and liability is gambling with your financial future. However, understanding insurance and risk is difficult. Most people do not understand risk management and the ability to insure against such losses. Managing risk is a major issue for businesses today. Often, snowplowers do not understand the risks they may incur. Unfortunately, a lot of the insurance companies today also do not understand the snow industry, thus making it doubly hard to adequately insure against loss and risk in our business.

WHAT IS RISK?

The term risk refers to the chance of loss, the degree of probability of loss and the amount of possible loss. There are two different kinds of risk. Speculative risk involves a chance of either profit or loss. It includes the chance a business takes to make extra money by purchasing new equipment, trucks and/or new snowplows. This also includes making business decisions in which the probability of loss may be relatively low and the amount of the loss can be projected. Entering the snowplowing business is an example of speculative risk because it requires some capital investment and may result in a profit or loss.

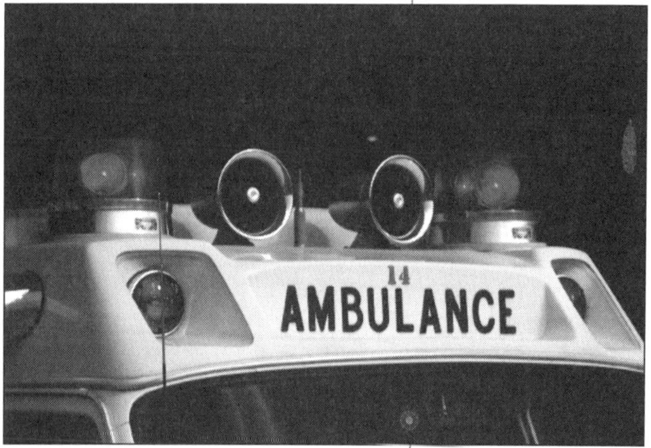

Snow removal professionals need to protect themselves from liability associated with the risks inherent in snow removal.

Pure risk is the threat of loss with no chance for profit. Pure risk involves the threat of fire, slip and fall, accident, or loss. If these events occur, some entity

Inspection of equipment before plowing can help prevent accidents.

incurs a monetary loss; but if the events do not occur, your company gains nothing.

The risk most snow professionals are concerned with, and wish to insure against, is pure risk.

RISKY BUSINESS

Once risks are determined and identified, companies have some options. These options include: reduce the risk, avoid the risk, self-insure against such risk or purchase insurance against the risk.

Reducing Risk. Reducing risk includes establishing some loss prevention programs. Such programs can include safety meetings, fire drills or plans on how to evacuate a building in the event of fire, lock out/tag out programs for equipment, mandatory seat belt use, ongoing training of employees, and other accident prevention programs. Other tips include:

- Have a written plan and map for each account, indicating where trouble spots are, pointing out heavily traveled areas, designating where to put snow piles, etc.
- Look for expected problem areas (poor lighting, low spots, potholes, etc.) and communicate these to the property owner. Put repairs needed in writing and take photographs of the problem areas.
- Conduct background checks on employees, particularly their driving records. A motor vehicle record gives you a history of accidents and violations. Don't assume that every person driving your equipment has a clean driving record, or even a license.
- Train and educate employees thoroughly. Develop standard operating procedures, make sure employees can safely operate necessary equipment, etc. Most accidents will come from your personnel, not you, so it's vital to make sure your staff has the same level of expertise regarding accident prevention.
- Inspect sites after work has been completed to ensure quality control.
- Make preventive maintenance of equipment a priority. Improperly maintained equipment can cause accidents by direct damage, or indirectly through rushed (and thus sloppy) work with a depleted fleet of equipment.
- Keep thorough and specific documentation for each account. Record location, start and finish time, crew members, weather conditions, lot conditions, etc.

Avoiding Risk. Avoiding risk is difficult. The most effective way (in the snow industry) to avoid risk is to not take on projects that have inherent risks associated with them. Examples of ways to avoid some possible risks in our industry would be to not take on projects that have a high probability of accident. They include:

- Plowing gas stations incurs high risk because of fill caps being slightly higher than the pavement surface.
- Parking lots with a high number of imperfections or manhole covers that stand higher than the pavement surface, and gravel parking lots.
- Sidewalk work because of the pedestrian traffic inherent to sidewalks.

INSURANCE

Reducing or avoiding risk is great, but we all know that running a profitable business involves some level of risk that is out of the control of the business owner, especially in the litigious society in which we live. Mistakes can and do happen, thus necessitating the use of insurance coverage.

An insurance policy is a written contract between the insured and an insurance company that promises to pay for all, or part of a loss. The premium is the cost of the policy coverage to the insured. Another explanation of premium is that this is the fee charged by the insurance carrier/company to cover specified losses that may occur.

Insurance companies are just like any other business. They are in business to make a profit. To assure themselves they will make a profit, the insurance company gathers information and data to determine the extent of the risk.

The reason insurance companies are able to assume the risk associated with a particular policy is known as the law of large numbers. This law of large numbers is the principle that states that if a large number of people (or companies) are exposed to the same risk, a predictable number of losses will occur during a given time period. Another term that is important here is the rule of indemnity. This is an insurance restriction that states an insured person (or company) cannot collect more than

SNOW TERM: SLEET

Raindrops that freeze into ice pellets before reaching the ground. Sleet usually bounces when hitting a surface and does not stick to objects. However, it can accumulate like snow and cause a hazard to motorists.[1]

Site inspections can reveal hazardous conditions, such as this raised manhole cover.

INSURANCE BASICS

The what, when and why of insurance coverage for snow and ice removal professionals.

Someone once said all politics are local. He could have just as easily been talking about insurance for snow removal contractors. The type and extent of coverage needed, and as a result the price, can vary greatly from one region to the next and from one company to the next. Are you a lawn contractor who does snow as incidental, off-season work? Chances are you'll pay less. Or, are you a large operation with snow removal making up a significant percentage of your business? Be prepared to write a fatter check than the guy down the street.

From the contractor's perspective, insurance is a complex, sometimes frustrating proposition. But, the alternative is a nightmare. In the litigious nature of our modern age, especially in metropolitan areas, even a minor incident can put a contractor who lacks the appropriate insurance coverages out of business.

Ultimately, the best plan for your company is finding a reliable insurance broker with a proven track record of working with contractors.

BASIC COVERAGES. A variety of insurance lines and programs exist, the selection and mix of which ultimately depend on each individual operation. However, in general, a snow contractor should consider the following coverages:

- **AUTOMOBILE:** Covers damages inflicted by or to a vehicle.

- **GENERAL COMMERCIAL LIABILITY:** Covers your operations (on-site activities) and the results of your operations, such as a slip and fall accident at a facility where you have conducted snow removal. Typically the maximum coverage of general liability is $1 million per occurrence, $2 million aggregate, so often an additional Excess Liability or Umbrella policy is needed.

- **UMBRELLA:** Provides additional coverage across all insurance lines, such as automobile, liability, equipment, etc.

- **EXCESS LIABILITY:** Provides additional coverage above and beyond General Commercial Liability. A lower-cost alternative to an Umbrella policy.

- **MOBILE EQUIPMENT:** Varies in name and specifications depending on the carrier, but covers equipment not covered through automobile insurance. It covers equipment not registered for travel over road, such as A-loaders, skid steers, etc.

Pricing for coverage can vary greatly from region to region and from one business to the next. A typical contractor in a rural area may get premiums for under $1,000 a year for general liability. A massive contractor who's doing $500,000 in payroll, may pay more than $10,000.

Further, if your business is solely snow removal, you will generally pay more than a contractor who does snow removal as an add-on service. The latter is cheaper because the coverages needed can be lumped into an existing liability policy at a lower rate.

A mobile equipment rider/policy can cover equipment not covered through automobile insurance.

SUBCONTRACTORS. A contractor should have a certificate of insurance from the subcontractor, and it would be for automobile, general liability and workers' compensation. Admittedly, finding subcontractors with general liability insurance is a challenge. One solution is to create short-term policies for your subcontractors.

CONCRETE CONTRACTS. Snow removal professionals can go a long way to reduce their potential liability, and thus an accident-related premium or deductible increase, by making sure the contracts they sign don't place undue burden on them. Unfortunately, many snow and ice removal contractors unknowingly assume liability by signing "hold harmless and indemnity" agreements with property owners/managers. Hold harmless and indemnity agreements may make the snow contractor liable for any damage or injury beyond his control. These agreements should apply only to the service being contracted and allow for a reasonable response period. These can put excess and unfair responsibility on the contractor. For example, if someone slips *inside* a building as a result of water or slush pedestrians have dragged inside, under a blanket hold harmless agreement, the contractor could be held liable. ❄

Reprinted with permission of Snow Business *magazine.*

Even the most careful snow contractors can have accidents.

the actual loss from an insurable risk. In other words, you cannot gain from risk management, you can only minimize losses. An example of this is: A company cannot buy two insurance policies and collect from both of them on the same loss.

POLICY TYPES

There are several types of coverage insurance companies will insure against. These types are:

- Property losses
- Liability losses
- Health insurance
- Life insurance

Eventually a snowplowing company will have to deal with all four. Property losses result from fire, accident, theft or other identifiable perils. Liability losses result from property damage or injuries suffered by others for which the policyholder is held responsible. Other types of insurance that we must deal with include workers' compensation, unemployment compensation, and Social Security.

Workers' compensation provides compensation for workers injured on the job. This includes payment of wages, medical care, job placement and, in some cases, vocational rehabilitation.

Unemployment compensation provides financial benefits, job counseling and placement services for persons who become unemployed through no fault of their own. Social Security provides retirement benefits, life insurance, health insurance and disability income insurance for the general population.

Snowplow contractors also

have to deal with fire insurance on buildings used for operating a business and automobile insurance on vehicles used in the operation of the business.

Other types that may come into play (both professionally and personally) include homeowner's insurance, business interruption insurance and professional liability insurance (if you are dispensing advice in a professional context).

WORKERS' COMPENSATION

Eventually, as your snowplowing business grows, you will begin to hire people to work with you. There are two types of people who will ultimately work for you in the growth of your plowing business, subcontractors and employees. We shall deal with subcontractors in a later chapter. For the purposes of this chapter, we shall deal with in-house employees.

Workers' compensation claims can put a company out of business if not taken seriously. While prevention is the preferred method of dealing with potential employee injuries, unfortunately such injuries do occur and must be dealt with.

Getting injured employees proper, timely medical care so that they can return to work quickly is a key factor in keeping workers' compensation premiums within reason. Not paying attention to what is happening can prove disastrous in the form of increased premiums, decreased productivity in your work force, and a decided effect to your bottom line.

Additionally, putting one's head in the sand and assuming the insurance company will "take care of it" is archaic and implies ignorance of the facts surrounding injury management.

And, make no mistake – the process should be viewed as injury management, as you can manage the process of getting an employee the proper care in an effort to return that employee to an active work capacity. This in no way implies companies should take a callous attitude toward good medical attention for an injured employee.

However, not paying attention to the proper aftercare may promote abuses to the system and, therefore, dramatically escalate the costs associated with such insurance coverage.

Document Thoroughly. When an employee is injured in any fashion, a report of the circumstance surrounding the work-related injury should be generated. Even minor injuries may balloon into major medical crises if not treated properly. If an employee fails to report such an injury, it can cause procedural problems later on if no one at your company is aware of the incident.

NOTABLE SNOWSTORM: March 1888: The "Blizzard of '88" produced temperatures plummeting well below zero degrees F, ravaging gusts of wind and deep snow drifts that stranded several cities, leaving them without transportation or communication. New York City suffered the most damage, particularly to its harbor areas.[2]

One person (an Injury Manager) should be in charge of documenting and following the injury from initial report to final disposition. All communication, documentation and information should flow through this one person so nothing is lost. This includes proper, regular ongoing communication with the injured employee, the adjuster in charge of the case, the medical professional and the company itself.

Your Medical Team. Having a panel of providers is essential to the entire process. Employees should be notified, when hired, that there is an approved panel of medical providers they are to see in the event of a work-related injury. While insurance carriers have a recommended panel of physicians, hospitals and medical facilities they work well with, the insured (your company) should be able to have input into what doctors/facilities are listed on the panel.

These providers have the same goal as the company. That is to be certain the employee receives the proper care and treatment and returns to work as soon as possible after recovery.

Should the work-related injury require hospital or medical attention, the person in charge of injury management has the right to, and should, accompany the employee to the doctor's office, hospital or medical facility. Some companies will insist the injury manager be with the injured employee as soon as possible after the injury occurs and then stay with the injured employee through the first hours of treatment, whether it is at a doctor's office, the hospital or the physical rehabilitation center.

Oftentimes, the injury is such (a non-emergency) that it can be treated in a doctor's office instead of a hospital, thus saving considerable dollars for the company or the insurance carrier. A good injury manager can assist in this initial determination.

Once at the doctor's office (or hospital outpatient facility) should the medical professional decide additional treatment is necessary, the injured employee and the injury manager can be transported to the recommended facility. The injury manager should then communicate with the doctor in charge to ascertain what additional immediate treatment is needed (if necessary) as well as what aftercare may be required.

Aiding Recovery. Unfortunately, some injured employees may not adhere to the prescribed follow-up procedures and/or appointments, thus causing inadequate or insufficient attention to proper ongoing care. The injury manager should stay on top of the situation by communicating with the injured employee regu-

SNOW FACT: The commonly used 10-to-1 ratio of snowfall to water content is a myth for much of the United States. The ratio varies from as low as 100-to-1, to as high as 3-to-1, depending on the meteorological conditions associated with the snowfall.[3]

larly. In some cases it is advisable for the injury manager to accompany the employee to the aftercare doctor visits to ensure the employee is adhering to the proper care program.

Keep in mind that, essentially, you are footing the bill for the employee's care through your insurance premiums, thus necessitating your participation in getting the employee healed and able to return to work. At these aftercare visits, the injury manager has the right to talk with the physician to ascertain what is needed in order to assist with the recuperating process.

Often, the case manager from the insurance carrier will also attend these appointments to keep accurate records of the entire process. Good communication is a key element in getting the employee back to work as soon as possible.

Getting Back To Work. Part of the injury management process is to have a light-duty work program in place so the injured employee can go back to work, earn a paycheck and feel productive. Additionally, during this time of light-duty work, the injury manager can oversee the remainder of the recuperation process. Sometimes this light-duty work schedule must be tailored to the individual employee's immediate injury needs; however, prior extensive studies show that getting him back to work is much more advisable than allowing the injured employee to sit around.

Insurance carriers typically have an approved panel of physicians to evaluate workers' compensation claims.

Interestingly, it is a proven fact that performing light-duty work often accelerates the desire to return to regular work status and positively affects the recuperation process. Unfortunately, in the middle of winter, this can prove a daunting task if there is no other work than snow work. A larger company can often find light-duty work within its core business base, but if you are just plowing snow it will be much more difficult to achieve an adequate light-duty, back-to-work program.

A huge factor in getting an injured employee back to work is communication. Communicate with the employee immediately after the injury occurs. Communicate with the medical professional to determine a

DEFENSE PREPARATION

When accidents happen, make sure you are prepared to tell your side of the story.

Accidents happen, and it is probable that at least once in your career as a snow and ice management professional you will have to defend yourself from an accident-related lawsuit. There are many things a company can do to help defend against such claims.

For example, with deicing contract customers, make sure you are proactively advising clients on pending weather and keep proof of this correspondence (copy of fax documents, phone log, e-mail, etc.).

Accurate, diligent record keeping and proactive advisement procedures in addition to the contractual agreements that shift the financial responsibility of a defense can help substantially minimize your involvement following a slip and fall accident. It is wise to realize that not only are slip and fall allegations costly to settle and costly to defend, but they are also costly in terms of time and anguish. Getting a third-party lawyer off your back as quickly as possible helps you get back to your business. Immediately upon becoming aware of a slip and fall accident, you should advise your insurance agent with the following details:

VICTIM:

- Name, address and phone number
- Where was he going? Where was he coming from?
- Why did he choose his particular path?
- What was his footwear?
- Was he carrying anything?
- Was he walking or running?
- What were the weather conditions at the time?
- Is there anything about him that may have caused the fall?

WITNESSES:

- Names, addresses and phone numbers
- What did they see happen?
- Where were they when they saw the fall? How far away?
- Did they see what type of footwear the victim was wearing?
- Was the victim carrying anything?
- Was the victim walking or running?
- What were the weather conditions at the time?
- Was there another, safer path?

- How could the victim have altered his or her behavior to avoid the incident?

OWNER/TENANT:
- When did he find out about the fall?
- Did he see what happened?
- When was it reported to him?
- What did he do for the victim?
- Does he know any of the witnesses?
- What were the weather conditions at the time?
- Exactly where on the property did this occur? Obtain an above-view drawing and take pictures of the area.

EMPLOYEES:
- When was the last time service was performed at the site?
- What was done?
- What were the weather conditions at the time? What were the conditions of the site?
- Were any hazards present (i.e., parked/abandoned cars)?

OTHER:
- Obtain statements from other contractors regarding the level of service they provided for their nearby customers.
- Obtain details of weather conditions from the nearest weather recording location. ❄

Courtesy of Scott McEachern, program manager, Greensure, Oshawa, Ontario.

Injured workers might be able to perform light-duty tasks after first returning to work.

proper care and light-duty work regimen.

Communicate with the insurance company's injury management specialist so he doesn't just keep writing checks to the employee for an unlimited amount of time if, in fact, that employee is able and medically ready to return to work.

Continue to communicate with the injured employee so he knows you care and that the company is concerned that proper care is given by the medical professional, and so you can keep tabs on progress with the recuperation process.

A successful businessperson will recognize the involvement of the company in the process is the key to keeping workers' compensation insurance premiums down and helping your employee return to health.

COMPREHENSIVE GENERAL LIABILITY

Snowplow contractors and all subcontractors should be covered for comprehensive general liability insurance claims. In a vast majority of the markets that receive snow, vehicles must have insurance coverage to be driven on the state and municipal roadways. However, this coverage isn't usually enough to adequately cover potential losses that can occur during plowing operations.

Commercial vehicle coverage will take care of claims associated with an accident that directly involves the driver and owner of the vehicle. If you slide through the plate glass window at Sears in the local mall, your vehicular policy (with commercial coverage) will take care of that problem.

However, should an accident occur after you have left the site (a slip and fall, or one vehicle slides into another on some patchy snow that remains on the surface after you have completed plowing operations), you would need a comprehensive general liability policy to assist in defending and satisfying a claim that may arise from such actions.

Simply stating to your customer that you are not responsible for any such claims is not enough to keep someone from filing suit against you or naming your company as a party to such an action. While you may not be found liable in such an action, you still need an attorney to defend you. If you have the coverage, the insurance company will pay to defend you if you are named as a party in such a suit.

Customers will regularly request that they be listed as "additional insured" on the certificate of insurance that is supplied to the customer. This is normal and customary. By doing this, the customer is asking that you defend them in the unlikely event that your actions (or inaction) while on the site are deemed negligent.

You should, however, read exactly what is being requested. Having a customer listed as additional insured is fine. Being listed on your policy as "named additional insured" is very, very dangerous. This wording means that your carrier may be asked to participate in any suit that names the customer.

FINDING AN INSURANCE BROKER

Because insurance can be a complicated matter, take care in selecting your broker. In searching for the right agent to handle all your insurance needs, interview them as you would any potential partner in your business. It is true that the right insurance carrier is indeed a partner in your success or failure in business.

Further, network with snow professionals you respect and ask them about their insurance coverages, their agent and their satisfaction with both. Such time-tested advice can be invaluable and most contractors are willing to help you avoid the mistakes they made.

BONDING

Should there be a need to supply bonds for a quotation you are preparing, your agent can assist you in securing this type of insurance also. Bonding a job is an insurance policy against failure to perform to the expectations of the customer.

Be very, very careful if a performance bond is requested for a snowplowing project. Performance bonds are subject to someone ascertaining if you have adequately performed your duties under the terms of the contract. Performance in snow removal is very subjective and wide open to interpretation. ❄

SOURCES

Snowplowing Operations, Loss Prevention and Claim Defense Tips, CNA Commercial Insurance. For more information on CNA, visit www.cna.com.

Watch Your Step!, presented by Scott McEachern, SIMA Snow & Ice Symposium, June 7-9, 2001, Denver, Colo.

[1]National Climatic Data Center, National Oceanic and Atmospheric Administration, AccuWeather.com, www.accuweather.com, National Snow & Ice Data Center, University of Colorado, Boulder, Colo.

[2]National Snow & Ice Data Center, University of Colorado.

[3]U.S. National Weather Service.

CHAPTER SEVEN: Pricing Snow Work Profitably

IN THIS CHAPTER

- ❄ Competing on service not price
- ❄ Pricing structures to consider
- ❄ Variables for contemplation

One of the most often asked questions regarding the snow and ice removal business is how to properly determine pricing. Pricing the services properly for your market, or circumstances, must be a high priority. When snow contractors get together in symposiums or at the coffee shop after a storm, pricing is usually a hot topic. Unfortunately, some companies think snow removal services are a necessary evil instead of a highly profitable revenue source. Contractors with this negative attitude are often pricing their service incorrectly or unprofitably. There are countless ways to price snow and ice removal services, but most are variations of one of four types of pricing structures: Per Push (or also called Per Occurrence), Per Season, Per Hour/Per Piece of Equipment and Per Inch.

A NOTE ABOUT PRICING

No matter what method is used to generate revenue for services rendered in winter, pricing should be one of the last areas used by customers in comparing potential service providers. Unfortunately, too many contractors use pricing as a primary sales tool, thus devaluing the industry as a whole and creating a market that competes on price, not service. Pricing our services must be dealt with in a manner that is consistent with our profit goals and the needs of our customers.

Customers who call you to request "hourly rates" for snow removal may not be asking the question the right way, or do not realize what information they seek. For example: if a potential customer calls requesting, "What is your hourly rate?" one possible answer is $165 per hour. When the customer states

Per-inch pricing is usually reserved for very large accounts such as airports and universities.

this seems high, one reply might be "How about $25 per hour?" The customer will probably like that pricing.

When asked how you will do the work you might reply, "At $165 an hour I'll send a payloader with a 14-foot scoop blade and do the job in 15 minutes. At $25 per hour, I'll send 15 guys with shovels and they will do the job in 4 hours." Now ask yourself, "What does the customer really want to know?" In truth, the customer wants to know how much you will charge for clearing his lot, not necessarily what the hourly rate is for your equipment. My point is that part of the "pricing question" is to properly ascertain what the customer seeks and communicate your prices in a manner that makes sense to the customer.

PER-INCH PRICING

Pricing on a per-inch basis is usually reserved for very large accounts in areas of highly variable snowfalls. Universities and airports are prime examples of where per-inch contracts are normally used. Further, customers who request such types of pricing normally have snow removal budgets in excess of $250,000, and often figure on spending several million dollars on snow and ice management in a given winter season. Quoting prices for such accounts requires a tremendous amount of knowledge on a variety of issues. These include accurate production times for all pieces of equipment and manpower that might be used, first-hand knowledge of the type of snowfalls that might occur at the site, probable moisture content of the accumulated snowfalls and prevailing wind direction of the probable snow event.

Pricing snow removal on a per-inch basis requires considerable expertise and knowledge of the intricacies of the account and the performance expectations. Per-inch pricing is normally not for the inexperienced plowing contractor with only a few pieces of equipment. These types of customers often require liquidated damages if the required equipment is not available during any given snow event. If you do not know your business and this industry inside and out, this type of pricing can put you out of business fast.

PER-PUSH PRICING

Per-push pricing is a structure in which you bill the customer a consistent rate each time you visit their facility to plow snow. Plowing contracts billed on a per-push basis should be the most profitable. However, be advised that pricing on a per-push basis requires considerable expertise, as you must know what your equipment (or subcontractors') production capabilities are in order to properly project revenue generation. Four-to-one and five-to-one ratios are common when comparing revenues to costs. And, an experienced contractor can project accurate per-push costs on sites as large as 150 acres of paved surface. Pricing projects on a per-push also allows customers to know exactly what to expect to pay when it snows.

It should be noted that intelligent and responsible contractors have a clause in their per-push contracts that allow for additional charges in the event snow accumulation exceeds a certain amount.

Additionally, if the contractor has to plow a particular site three or four times during the snow event, he should set up the contract so additional revenue is generated for each separate visit to the site. In these cases it is recommended the contract with the customer allow the contractor to make the decision as to when to plow and apply deicing material.

There should also be a clause that advises customers "plowing and/or salting may not reduce the lot to bare pavement and snow or ice accumulations are naturally occurring events that the contractor should not be held responsible for."

PER-HOUR PRICING

Pricing per hour is a structure in which the customer is billed based on the time you spend at their facility. Pricing per-hour/per-truck is the easiest way to avoid learning about snowplowing as a business. Such methods of pricing allow the guy next door to get into the plowing business. Little, if any, expertise is required. Errors in judgment are the fault of the customer and not the contractor.

The margins are generally much lower, as contractors have a tendency to price to compete

SNOW TERM: SNOW DENSITY
The mass of snow, per unit volume, which is equal to the water content of snow divided by its depth.[1]

PRICING STEP BY STEP

The steps of pricing a snow removal job.

Everyone is searching for the cure-all to pricing, a one-size-fits-all method to set prices. Unfortunately, such models do not exist. Snow removal is highly variable and, as a result, pricing methods can vary widely.

Pricing is not an exact science and many contractors rely on their experience to evaluate a property and set a price. Most likely, as you gain experience, you will develop your own method, which fits your market and your business. The key thing to remember is that you are offering a professional service and should charge accordingly. Competing on price and slashing fees to earn a job usually ends up hurting both you and the client. And, such low-ball pricing hurts the industry as a whole. Make sure you know your costs, add an appropriate profit level and then make a strong presentation to the customer.

Most successful contractors follow these basic steps:

1: VIEW THE SITE

Obviously to price a job accurately you must review the site. Pay special attention to any unique characteristics that could add time or liability to your snow removal operations.

2: EVALUATE THE CLIENT'S NEEDS

Some customers want bare pavement and clear sidewalks. Other customers just want a clear path to drive in and out on. Interview the client to get a sense of his demands and evaluate the time necessary to fufill those demands.

3: ESTIMATE THE REQUIRED TIME, EQUIPMENT, ETC.

Estimate the time, equipment, personnel, materials, etc., it will take to do the job to the customer's satisfaction.

4: CREATE A DOLLAR AMOUNT

Put a dollar amount to the time and equipment required.

5: ADD PROFIT

Add the desired level of profit to the overall expense. Four- or five-to-one revenue-to-expense ratios are common.

6: PRESENT SALES PITCH TO CUSTOMER

Make your best presentation. Justify the price with value. ❄

with the other guy who is pricing by the hour. Growth patterns are accelerated (although at decidedly lower profit margins than can be obtained by alternative pricing methods) because there isn't the need to visit every site that is quoted prior to adding it to your customer list.

There also needs to be a tremendous element of trust between the customer and the contractor who prices all of his work by the hour. Most contractors are honest and fair, however, some unscrupulous contractors can add "ghosts" to a jobsite in order to increase revenues. This is because the unsuspecting customer is not usually at the site at 3 a.m. to check how many pieces of equipment are being used. These contractors always get caught eventually, which makes it that much harder for the honest contractor to generate a trusting relationship with the customer. Many national accounts require per-truck/per-hour pricing because it is easier for bidding purposes and, in those cases, it is often easier to take the account than to attempt to re-educate the customer.

SEASONAL CONTRACTS

Seasonal contracts are those in which the customer is charged a consistent fee, typically by month, for the entire winter season regardless of the amount of snowfall. While such contracts typically include provisions for minimum and maximum snowfalls, they provide a consistent source of cash flow to the contractor.

To set seasonal rates, a contractor must determine the average number of events in their locale. An event is defined as any precipitation that requires a snow contractor to provide any service on a customer location.

A mix of pricing methods can prevent losses due to a mild or extreme winter.

These services include plowing, deicing, sidewalk clearing, pre-salting, etc. It's not an easy task to determine how many events your market area receives each winter. Some history is necessary for such a determination and a review of the weather occurrences for the past several years is required. Such information can be found on the Internet at various sites, including:

- www.millenniumweather.com
- www.weatherunderground.com
- www.nws.noaa.gov
- www.weathertap.com
- www.weatherwatchers.com

Other avenues to ascertain this information include the local library, local weather station meteorologists and local colleges or universities.

Once you have determined the number of events that occur, you have to determine how much work will be required to serve the client during a snow event. While there is not a set of all-encompassing rules to determine this required level of work, past experience will be a tremendous asset. Once this is determined and a value assigned to the total amount of work necessary for the winter season, this will allow the contractor to recover overhead costs associated with the reduced workload that winters normally bring.

Per-season pricing contracts are usually tied into other services, such as landscape maintenance, parking lot sweeping, or a complete grounds maintenance service agreement. This allows for a year-round contract that is all-inclusive of the summer and winter maintenance

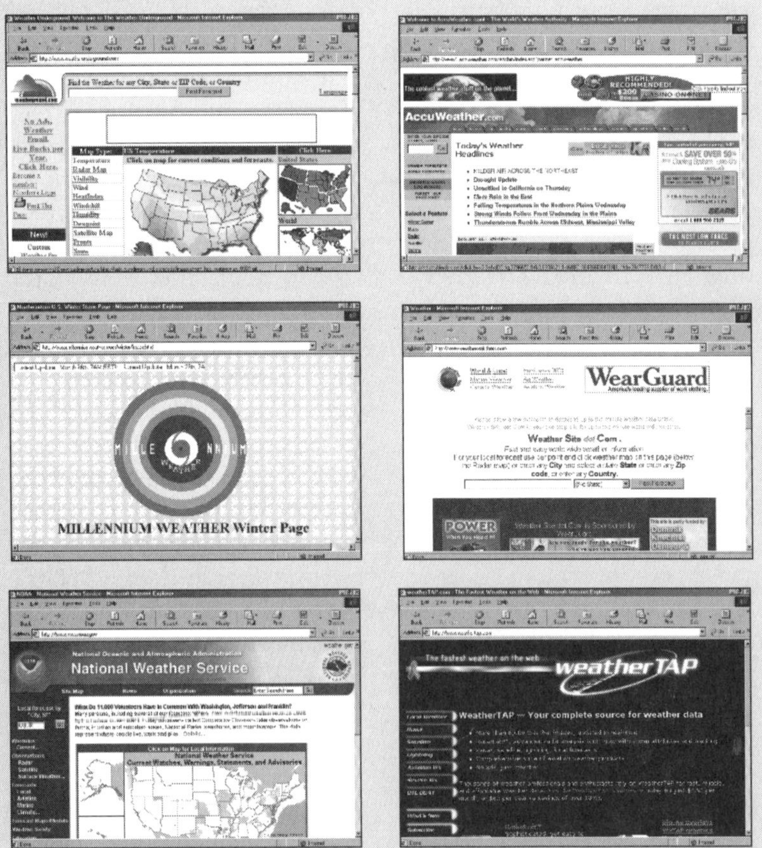

Weather forecasting Web sites, which typically feature historic weather almanacs, can be used to set seasonal rates. Such sites can help snow contractors determine the average number of winter events in their market.

services. If you know the average number of times you plow in a given season, you can project how many times you will have to bring out the equipment during the winter season.

Taking on seasonal work for only one season can be disastrous. Someone almost always loses with a one-year contract. With a three-year contract both the contractor and the customer can assume the law of averages will make it all come out even in the end. In those years when there is little snowfall, the contractor knows how much income will be derived through the monthly payments made by the customer.

In those years when there are above-average snowfall totals, the contractor might lose money on this particular account. However, if the other types of pricing snow services are also used (on other accounts), then the shortfall is generally made up due to the dramatically increased revenues generated by per-push and per-hour/per-truck pricing strategies employed with some clients.

MIXED PRICING

If all the plowing contractor's business is priced per-push or pe-hour/per truck, and there is a mild winter, revenues drop below anticipated levels and cash flow problems can become prevalent. Adding a mix of per-season customers can avoid these kinds of pitfalls.

For contractors in most markets with snow revenues less than $1 million a year from snowplowing, pricing structures normally fall in the per-push, per-hour/per-truck, or per-season categories. A mixture of these three pricing structures is preferred, allowing the contractor to maximize the benefits of each pricing structure, while minimizing the downside of a mild winter season.

If a contractor can project what revenues he needs to survive a mild winter (these numbers should be readily available if you are doing your cost accounting properly), then securing enough per-season work allows the responsible contractor to guarantee adequate cash flow in winter. Then, once this cash flow need is satisfied, remaining customers can be billed on a per-push or per-hour basis.

OTHER CONSIDERATIONS

Some contractors use a non-refundable retainer factor or minimum billing method of securing revenue prior to the start of the winter season. Customers are then billed against the retainer for services rendered. This is an efficient way of projecting cash flow. If the retainer is exceeded (because of above-average snowfalls) the customer is invoiced for the additional services rendered at an agreed-upon rate. A large number of contractors in mar-

SNOW FACT: Most snowflakes are less than a half-inch across; however, they can get as big as 2 inches across. In 1971 a snowflake measuring 8 inches by 12 inches was recorded in Bratsk, Siberia.[2]

SNOW SERVICES SPECIFICS

Once the snow starts to fall, contractors have to be prepared for the onslaught of calls from residential and commercial customers who desperately need to be shoveled out of the driveway or have a path cleared to their businesses.

When the rush hits, contractors have to have their pricing set.

THE PRICE POINT. Contractors have several options when pricing their services.

"For commercial properties, we price by square footage," stated Jay Goldberg, president of Cedar Glen Property Maintenance, Burlington, Vt. "For residential properties, we have a standard driveway price that we usually use. If the size varies from job to job, we will price each job by the hours worked on the job."

Similarly, Mark Shearholdt, president of Certified Maintenance Services, Manchester, N.H., said that he prices jobs based on the time spent on them.

"We base our pricing on our rate for each minute," commented Rick Kier, president of Pro Scapes, Syracuse, N.Y., "I place a dollar value on each minute. We look at the job and evaluate how many minutes the job will take."

Kier continued by explaining his pricing structure. He charges $165 per hour for a standard truck. This comes out to $2.75 for each minute.

"We have prime-time hours – 3 a.m. to 7 a.m. – and off-time hours with different prices," Kier added. "If the job doesn't have to be completed until 7:30 a.m., we charge $137 per hour. If we have until 8 a.m. or later, we charge $108 per hour."

Alan Steiman, CEO of Alan Steiman's Landscape, Northboro, Mass., explained that there are four ways to charge for snow removal services.

- Price by hour using the cost of the equipment. According to Steiman, this is the least profitable because the contractor can't make more money than he originally agreed to.
- Price per inch of snow. This requires knowing how the production cost figures into every inch of snow on a property.
- Price per push. A flat charge each time the plow is pushed.
- Price seasonally. According to Steiman, determine a price

by the average number of snowfalls each year. This may not be the most profitable in a year with a lot of snowfalls, but during a year with less snowfall, this pricing structure can be very profitable.

RESIDENTIAL VS. COMMERCIAL. Snow removal contractors target residential customers, commercial customers or sometimes both, and there are several differences between the two.

"Residential properties require a lot more attention to detail," Kier noted. "With residential, we need a good system of keeping track of details. For example, we keep notes on where customers want the snow piled. This information is stored in our computer so whoever is plowing can see the notes for each property each time they plow."

Kier explained that he prefers commercial customers because the work is not as demanding and there is less paperwork because there is less billing per dollar.

"Residential is much easier and must less of a challenge," Goldberg countered. "Residential properties usually need plowed by 7 a.m. and then we don't have to go back until 5 p.m. to get the snow cleared out again. With commercial, the business usually has to stay open all during the day."

Goldberg explained that the clients determine which type of property contractors generally prefer. "We target customers who want the job done no matter the cost," he explained. "These are easy customers to work with because they just want the snow gone. Our No. 1 clients are companies that have freights coming in and out all day long. They want us to do whatever we can to keep it clear."

A challenge with any property, however, is dealing with sidewalks. "For sidewalks, we charge $32 based on one man and a shovel," stated Kier. "It is important to know the average prices in your area. Our rate is high enough for a profit and to pay well."

"We try to avoid properties with sidewalks or we try to get the owners to take care of them," Goldberg added. "We don't include sidewalks in the contracts, then we offer a separate price for the sidewalk and they have to pay for the service each time. A lot of people won't pay this per-time price so they will do it themselves or hire someone specifically for the sidewalk work."

"The key to success in this industry is planning ahead," Goldberg continued. "By Aug. 1, we have all of our snowplowing services sold for the coming winter." ❄

Reprinted with permission from Lawn & Landscape *magazine.*

kets that get little snow have gone to this pricing strategy as a way to generate significant revenues during slow snow years.

No matter what method (or combination of methods) you select, be aware that snowplowing is a viable profit center no matter how little or how much snow your geographical area receives. Making a profit at snow and related snow services is a mindset. As long as it is thought of as a profit center — like landscape installation, landscape maintenance, chemical lawn care, excavation or power sweeping — money (good money) can be made from snow and ice management services.

NOTABLE SNOWSTORM: **January 1918: A tremendous blizzard shut down the Midwest, stopping mail service for two weeks.**[3]

VARIABLES TO CONSIDER

Even though there are market and weather variables unique to each region, marketing, pricing and sales operate much the same in Peoria as they do in Salt Lake City, even though Peoria receives wet and heavy snow and Salt Lake City receives dry and fluffy snow. This is not to say there are no variables between markets. There are. In fact, the variables can be significant from region to region.

I often hear arguments that one particular region of the country is mired in per-hour/per-equipment pricing structures and per-push pricing structures will not be accepted. That notion is a falsehood.

Two immediate things come to mind. Many said mass-producing motorcars was impossible because there could be no consistency in the parts manufacturing. This was proven wrong. And, at one point, many people said one plowing contractor could not possibly manage to plow snow commercially in an area he did not live in. In both cases, challenges did enter the equation, but those challenges were overcome with intelligent use of all the information available to these people – and by thinking outside the box.

Yes, chances needed to be taken, but these chances were more than dumb luck. They were well thought out and implemented with care.

Let's take an example. If you live and work in a market where snow management services are driven by the per-hour/per-equipment model and you want to go outside that paradigm to increase your margins on the work accomplished, you might approach it the following way.

Take on all the snowplowing work you can at the per-hour/per-equipment pricing structure. Once you have filled up your book of business and think you cannot take on any new business, think twice before turning away the next call that comes into the office.

Go look at the project and tell the customer the only way you can take on any more work is at this per-push number. And, you require a two plowing deposit up front, which you will bill against as it snows. The customer may walk away and go elsewhere.

However, what did you have to lose by approaching this quotation in this "new" manner? Previously you were prepared to walk away from the customer's business anyway.

Most of the time, the last minute calls for a quotation are a result of poor planning on the part of the potential customer. He may be desperate for a contractor to take on his site. Maybe the incumbent contractor went out of business at the last minute. Who cares what the reason is? You really don't need the added work. So, you alter your thought process.

It may take five of these calls to get one to consider this option. But, eventually one customer will be forced to take your quotation seriously. That will be followed by another. And, as that happens, you'll be able to drop a difficult and unreasonable customer from your list, or you can drop that customer who takes 90 days to pay your invoice – in favor of this customer who is willing to pay a portion of his bill up front.

Many last-minute snow removal calls come from the residential sector.

There are contractors in markets who have taken this advice to heart, and after a few years have only customers who pay per-push and also pay some of the money up front. Creating this condition within your company may take a while, but patience is a virtue that often handsomely rewards the diligent.

If no one thought outside the box and tried to improve what they were doing, we'd all still be using two-piece phones in the parlor or the living room. ❄

SOURCES

[1]Webster's New World Dictionary.

[2]National Snow & Ice Data Center, University of Colorado.

[3]National Climatic Data Center, National Oceanic and Atmospheric Administration, National Snow & Ice Data Center, University of Colorado,Boulder, Colo., http://nsidc.org. AccuWeather.com, www.accuweather.com, .

CHAPTER EIGHT: Law, Taxes & Contracts

IN THIS CHAPTER

- Basic business laws and regulations
- Business tax fundamentals
- Snowplow contracts

The world has changed throughout the last 50 years. No longer is a handshake and our word good enough to get by when dealing with customers. Larger, national organizations that hire us to manage snow and ice on their sites simply will not do business without a written and fully executed contract in hand. This is the way business is conducted "out there." There is nothing to be afraid of when working with contracts. The laws that apply to these contracts are such that underhanded, one-sided contracts are rare today.

Commercial customers want their vendors to make a profit as they have come to realize that without profit a company (even a snow management company) cannot survive. No large retail facility wants to beat up its vendor to the point where it is cheaper for that vendor to go out of business rather than continue to service the account. What could be worse than Home Depot finding out its plowing contractor closed up shop just before a 12-inch snowfall hit the area? Do not, for a minute, think this is what Home Depot, WalMart or Joe's Deli want to have happen.

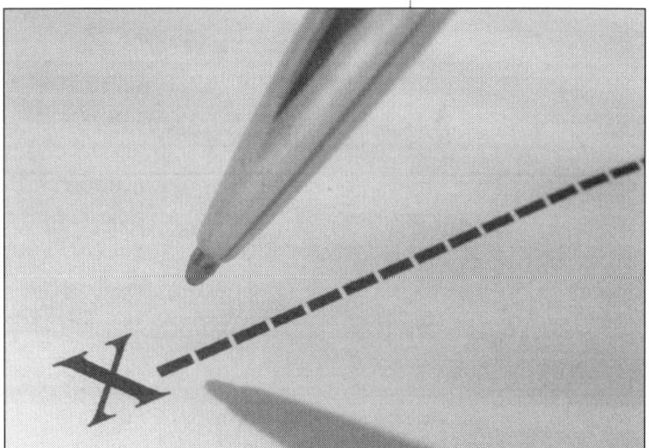

However, these people also assume we are all businessmen who know our costs and have quoted a viable price for the work we want to do for them. Contracts between customer and vendor are there to make certain all parties understand what is expected of each other. Very large

No longer is a handshake and your word good enough to get by in today's business world.

organizations use contracts that they have paid lawyers to write. These attorneys are working on behalf of the organization that hired them, so it is reasonable to believe the contracts are written in favor of their client. Wouldn't you, if you hired an attorney to write your contract, want the attorney to act in the best interest of you and your company?

The world of business is governed by laws and regulations.

LAWS & REGULATIONS

Laws are enacted to protect everyone, not just the big companies. Imagine the chaos if we had no laws to work with or from, no speed limits to control how we drive, no limitations on the medical profession, or no restrictions on taking things that don't belong to us. The world of business is also governed by laws. The trend during the 1990s seemed to be that government was stepping in more and more to govern the behavior of business people. Thus, we see more laws and regulations regarding sexual harassment, proper treatment of those with disabilities, proper treatment of employees with regard to hiring and firing, working with subcontractors, and enforcement of environmental and safety laws.

As one might expect, businesspeople often want to set their own standards. Unfortunately, there are a few unscrupulous people who spoil it for the rest of us. Also, at conventions for small business owners, you'll often hear people say that "this law doesn't apply to us" or, "I won't follow that principle of law." And, too often, the person just starting out in business will think that it is too expensive to follow the laws that apply and he cannot make a profit if he follows the law. These laws are implemented for the good of all concerned, and are there to guide us in how we conduct our business as it relates to others we have to interact with.

Business Law. Business law refers to rules, statutes, codes and regulations established to provide a legal framework within which business may be conducted and that are enforceable by court action. Business people today need to have some working knowledge of laws regarding sales, contracts, liability, taxes and bankruptcy. There are two major kinds of law: statutory law and common law. Statutory law (written law) includes state and federal constitutions, legislative enactments, treaties of the federal government, and ordinances (written laws). You can read these laws; however, it is often very difficult to determine their meaning. Unfortunately, this is why we have so many lawyers and courts.

Common Law. Common law is the body of law that comes from decisions handed down by judges. Common law is often referred to as unwritten law because it doesn't appear in any legislative enactments, treaties and so forth. Such decisions are called *precedent*, and they guide judges in handling of new cases.

Tort Law. Tort law relates to wrongful conduct that causes injury to another person's body, property or reputation. These types of laws come into play in the event of a slip and fall incident. Liability laws fall under Tort Law and hold a business liable for negligence in performance of its duties with regard to safety and execution. Thus, if you fail to adequately clear a property while plowing, you may be held liable for any injury that may occur as a result of your negligence in adequately clearing the lot.

TAXES

Tax laws can be very daunting. Often, tax issues lead to spirited discussions as to the merits of any one particular tax issue. Let's define the major taxes levied upon individuals and businesses today.

- Income taxes are those taxes paid on the income received by businesses and individuals. Income taxes are the largest source of revenue received by the U.S. Government.
- Property taxes are those taxes paid on real and personal property. Real property is real estate owned by individuals and businesses. Personal property is a broader category that includes any movable property such as tangible goods. Taxes are based on their assessed value.

NOTABLE SNOWSTORM: January 1922: The "Knickerbocker Storm" dumped more than two feet of heavy snow on Washington, D.C., causing the roof of the Knickerbocker Theatre to collapse, killing nearly 100 people.[1]

- Sales taxes are paid on merchandise when it is sold at the retail level.

The taxes we (as snow management professionals) need to be the most concerned with are income taxes (and the related taxes for employees) and sales taxes. We must pay income taxes on the profit we derive from the fruits of our labors in this industry. Some states require companies collect a sales tax on the services provided. These sales taxes must be paid to the corresponding state at the appropriate time.

Businesses are taxed in a variety of ways, and the assistance of a qualified accountant is advisable.

CONTRACTS

A contract is a legally enforceable agreement between two or more parties. A contract is legally binding if the following conditions are met:

- An offer is made.
- There is voluntary acceptance of the offer.
- Both parties must give consideration (something of value, such as money) in exchange for services rendered.
- Both parties of the contract must be competent.
- The contract must be legal (example: gambling losses are not legally collectable).
- The contract must be in proper form (example: an agreement for sale of goods or services worth $500 or more must be in writing).

A breach of contract occurs when one party violates the contract. Should a breach of contract occur, the following may happen: The person who violated the contract may be forced to live up to the agreement; payment of damages may be awarded to a person who is injured by the breach; restitution can occur whereby if the person fails to live up to his end of the contract, the other party could agree to drop the matter, and wouldn't have to live up to his part of the agreement either.

Snowplowing contracts vary widely from company to company and from region to region. However, there are some consistent elements. The following are the principal parts of a typical snowplowing contract.

Headnote. Each contract should feature your company logo and contract information at the top of the first page. This communicates clearly who the bidding company is and how customers can contact the firm. A professional logo and overall quality design of the document can also help create a positive image of the company.

Submitted To. A brief description/representation of to what company and which contact the contract is being submitted.

Description Of Services. Describes what the customer is receiving as part of the contract. This area can be quite simple, direct and to the point. A brief description of what is to take place, and when and how much this service will cost are all the pertinent details the customer will want to know up front.

Services Not Included. Some contractors also list related services that are not included as part of the contract and how those "other" services will be billed. For example, in parts of the country where heavy snows are not routine, a contractor might charge an extra fee when physical removal of snow with heavy equipment is required. This section provides an area for the contractor to clearly communicate these special services to the customer.

Insurance. Some contractors include language describing the insurance they hold and if the company is bonded.

Terms And Conditions. The Terms and Conditions is where the meat of the contract should lie. This is where you detail anything that deviates from the norm, or any special conditions the customer needs to be aware of prior to agreeing to have you service his property.

A three-part contract with these terms and conditions preprinted on the rear is an excellent way to make sure all parties know what is expected of everyone during the contract term.

At the bottom of the front of the proposal, where the customer signs to accept the agreement proposed, should be language that specifically states that the customer has read and accepted the stated terms on the rear of the contract document.

In the terms and conditions portion of your contract you should detail all conditions under which you are accepting responsibility for the services you are to provide to the prospective client. While you cannot abdicate all responsibility for anything that might occur during or after a snow/ice event, it is possible to shift some of the liabilities for accidents to another party (ostensibly, the owner of the site you are servicing). Language to that effect should be contained in the terms and conditions area of the contract.

One example of contract language in the terms and conditions that can help snow professionls limit their liability

SNOW TERM: WINTER STORM WARNING

Issued when a combination of heavy snow, heavy freezing rain, or heavy sleet is expected. Winter Storm Warnings are usually issued six to 24 hours before the event is expected to begin.[2]

Proposal & Contract

XXXXX COMPANIES

SNOW PLOWING DIVISION

XXX West Street, Erie, PA 16220

(814)XXX-XXXX FAX (814)XXX-XXXX

PROPOSAL SUBMITTED TO	PHONE	DATE
STREET	JOB NAME Same	
CITY, STATE AND ZIP	JOB LOCATION	

We hereby submit specifications and estimates for:

Snowplowing of driveway/parking lot areas at above location on per plow basis. Plowing to be done after 2-3" (prox) accumulation of snow as the result of snowfall or drifted snow.

Cost per plow $

Salting of above parking lot as needed with granular salt $ per occurrence

_______ Please check here if you wish XXXXX Companies to salt the parking lot at our discretion.

Cost to clear walks once per day after 1"-2" (prox) accumulation of snow. Per occurrence $ per time

Calcium Chloride application to walks as needed - $ per bag or any part thereof, plus the cost to clear walks.

Above cost does not include hauling or stacking snow, although these services are available if requested.

This pricing is guaranteed for 3 seasons and is valid thru the 2002-2003 winter seasons.

Contract cancellable upon written notification.

Payment to be as follows:

net 30 days

All material is guaranteed to be as specified. All work to be completed in a workmanlike manner according to standard practices. Any alteration or deviation from above specifications, involving extra costs will become an extra charge over and above the estimate. All agreements contingent upon strikes, accidents or delays beyond our control. XXX Companies not responsible for "Acts of God", and assumes no liability for naturally occuring conditions. Our employees are fully covered by Workman's Compensation Insurance.

Authorized Signature ____________________

Note: This proposal may be withdrawn by us if not accepted within 14 days.

Acceptance of Proposal - The above prices, specifications and conditions are satisfactory and are hereby accepted. You are authorized to do the work as specified. Payment will be made as outlined above. I have read the Terms and Conditions and understand them.

Signature ____________________

Signature ____________________

Date of Acceptance: ____________________

Example of a snow and ice removal contract.

TERMS AND CONDITIONS

This written agreement contains all conditions and describes all work to be done. This agreement supersedes all previous agreements and any verbal commitments made prior to date of this agreement.

"Per Plow" rate is for snow accumulations up to 7" (prox). Accumulations greater than approximately 7" may result in additional charges to compensate for additional time required to adequately clear snow from lot. Customer understands that plowers may have to go thru lot once to clear heavy snow and then re plow lot to clear remaining snow. Customer understands that this may result in additional charges for that particular snowfall. Customer agrees to pay such additional charges.

All plowings shall be considered "full plows" unless otherwise agreed upon by the parties hereto, prior to plowing being done. Partial plows may be charged at a prorated lower amount than that quoted -depending upon amount of work required. Quoted rate is for snow plowing of parking areas (or driveways) only. Unless specifically stated, sidewalk snow removal is not part of quoted rate. Quoted rate does not include salt spreading services or front end loader services, however these services are available at an additional charge.

Quoted rate is for plowing that is done as part of regular service when all other accounts are being checked or plowed. Special request plowing (for example -a phone request to plow a driveway or lot at mid-afternoon) -is subject to a minimum charge. This is due to the fact that our driver's get paid a one-hour minimum when called out for "special plowings". Customer to be advised of possibility of minimum charge at time of "special request".

The owner of the property is presumed to know their own property boundaries. The Owner will clearly stake the areas to be plowed. In the event the areas staked are erroneous, the owner agrees to defend and hold harmless the contractor for any and all trespasses or damage that may result from the owner's failure to properly stake his property.

xxxx Companies will exercise its best judgment based upon weather forecasts and existing conditions at the time. Customer is aware that weather conditions in the Erie area may change rapidly and without notice. Changes in weather conditions are considered to be an "Act of God" and xxxx Companies assumes no liability for Acts of God. Reports of damages must be reported to the contractor within forty-eight (48) hours. Failure to report damages constitutes a waiver and the contractor is released from liability.

Customer agrees to allow xxxx Companies to decide if snowplowing is warranted based upon snow accumulations at customers' particular location. Customer understands that snow accumulations may vary throughout the Erie area, and that accumulations in one section of Erie are not necessarily indicative of the accumulation at the customers' particular location. Customer also understands that drifting snow may necessitate plowing of their particular location, regardless of the total snowfall at that location. In any event customer agrees to allow xxxx Companies to decide if snowplowing is necessary.

Accounts that are past due will not be plowed until account is brought up-to-date. Customer understands and accepts the fact that the delays in payments made to xxxx Companies may result in appropriate legal action being taken to collect monies owed to J xxxx Companies. Customer understands and agrees that costs of such legal action, including without limitation lawyers fees, costs, and expenses of suit or bringing suit, may be passed on to the customer, and customer accepts this condition. Contract shall be binding inure to the benefit of the parties and their heirs, executors, administrators, and assigns.

Customer understands that plowing (or salting) of a particular location may not clear the area to "bare pavement" and that slippery conditions may continue to prevail even after plowing (or application of salt). Customer understands that xxxx Companies assumes no liability for this naturally occurring condition. Customer agrees to defend and hold harmless the contractor for any and all trespasses or suits that may arise as a result of this naturally occurring condition.

If sidewalk snow removal is selected as an option: Customer understands that sidewalk crews may not work safely if temperature and wind conditions combine to make Wind Chill factors below 0 degrees Fahrenheit. Customer agrees and understands that Allin Companies reserves the right to stop working In these severe conditions (without penalty) so as not to force unsafe conditions upon our employees.

This contract is cancelable upon written notification by either party. Monies invoiced or due for services rendered are due and payable upon such cancellation.

comes from Scott McEachern, program manager, Greensure, Oshawa, Ontario.

According to McEachern, contracts should be worded in a way that financial responsibility is not assumed when any of these three concerns are present:

1. In any and all cases in which the contractor is restricted by the property owner from sanding or salting at the contractor's own discretion;

2. In any and all cases in which the damages or injuries to persons or property or claims, actions, obligations, liabilities, costs, expenses and fees arise as the result of incidents occurring on areas of the property not serviced by the contractor;

3. In any and all cases in which the damages or injuries to persons or property or claims, actions, obligations, liabilities, costs, expenses and fees arise as the result of incidents occurring during times in which the contractor is restricted from accessing the property to perform his or her duties.

Keep In Mind. While a contract is a legal document that binds both parties to the agreement, you should be aware the court system generally treats residents differently than commercial/industrial/retail customers when interpreting the agreement in the unfortunate event of a dispute. Residents are deemed to be not as savvy about legal matters as are those who operate in the everyday business world. Thus, it may be important that the residential customer really does understand what he is signing. Putting the words about having read and accepted the terms and conditions in bold black type should be proof positive that you did not attempt to hide anything from an unsuspecting residential customer.

In contrast, commercial customers are often considered to be savvy about signing contracts, as this is part of the ongoing business duties of any good manager.

CUSTOMER-SUPPLIED CONTRACTS

As your company grows and begins serving larger, more sophisticated clients, you will find many of the contracts will be written by your client, who will ask you to sign the contract. These contracts usually are written by the customer's legal department. Often, these customers will ignore your written contract and ask that you sign their version.

This happens more often than one might think, and is standard practice with large clients who have multi-location sites to be serviced. Many times these sites are located in various parts of the country and are serviced by numerous independent snow contractors of various size and type.

It is recommended you read these contracts closely. Often they contain language that might scare or concern you. You should be concerned. While it may not be advisable to refuse to sign such a contract, you need to know what you are signing prior to doing so. Remember, these companies pay lawyers top dollar to write contracts to protect them. These contracts tend to be geared toward protecting the client, not you.

After you have carefully read the contract, the decision of whether or not to sign the contract essentially becomes a business decision. This decision is whether or not to accept their terms and conditions in lieu of your own. Sometimes these contracts can be negotiated, and sometimes the customer refuses to negotiate anything.

Sometimes the size of the contract will dictate whether negotiation can take place. And, if the customer has numerous bids of a like size and dollar amount, he will take the vendor who will agree to his terms and conditions. This, too, is a business decision that must be made on the part of your potential customer. However, one might be surprised as to just how much negotiation can take place. It never hurts to ask.

Using a written contract is good business and good practice. In this way everyone knows what is expected, what your company will do in most situations, and what the compensation (consideration) is for both parties. ❄

SNOW FACT: The fluffiest snow falls at temperatures around 15 degrees F.[3]

SOURCES

[1]National Climatic Data Center, National Snow & Ice Data Center, University of Colorado, Boulder, Colo., http://nsidc.org. AccuWeather.com, www.accuweather.com, National Oceanic and Atmospheric Administration.

[2]U.S. National Weather Service.

[3]Webster's New World Dictionary.

CHAPTER NINE: Financial Management

IN THIS CHAPTER

- ❄ Strategic planning
- ❄ Making a business plan
- ❄ Creating and using a budget
- ❄ Understanding financial statements
- ❄ Key ratios to watch
- ❄ Managing cash flow

Everyone knows that good financial management is essential to the success of any business entity, including a snow and ice control business. To be successful in your quest to grow, some information on finances is imperative.

A basic knowledge of budgeting, planning, cash management and reading a financial statement will go a long way to your ultimate success in the snow and ice removal industry. All too often we are concerned about getting the work done and we don't really pay attention to the finances. Successful entrepreneurs will tell you that if you can manage finances and if you operate "by the numbers," you can determine (in any business venture) if you are operating profitably. And make no mistake about it, you must be profitable to succeed.

Strategic plans, business plans, financing plans, forecasts and budgets all portend to do the

same thing: They assist you in planning and controlling your business. Basically, these are all terms for the same thing. Unfortunately, the business community, the financial community and even the government tend to misuse these terms. Most often they will select a term they

Strategic plans, business plans, financing plans, forecasts, budgets and financial statements assist you in managing your business.

are familiar with and then apply it across the board.

Let's define the various terms used in the business and financial communities. We will then specifically address those few that really do apply to what emerging snow and ice control companies need to know.

STRATEGIC PLAN

A strategic plan is a long-term plan. Normally, it looks five or more years into the future. This type of plan lays out a company's long-term objectives in narrative fashion, covering all major activities — markets, services, finances, human resources and (in our case) subcontractor use. This narrative might be supported by detailed financial forecasts of balance sheets, income statements and cash flow schedules for each year of the plan. For our purposes herein, we shall not discuss strategic plans in depth as these are usually reserved for large, multimillion dollar entities that may have to deal with a Board of Directors or stockholders.

Nonetheless, it is never a bad idea for an owner of a snow and ice removal company to think long term and to set business goals that are five or more years out.

BUSINESS PLAN

A business plan covers a shorter term than the strategic plan. This may encompass one or two years. Often the business plan will also function as a basis for financing plans given to a bank when applying for loans. This plan should serve as the working tool for the owner to use to develop tactical plans for monitoring the performance and development of the objectives to be used in growth strategies. If used internally, this plan may be short in nature and succinct in its detail. However, when used as a financial tool, the plan can be quite lengthy. In this instance, the business plan would include background of the company (or the company founder if a new entity), a marketing plan, operating plan, staffing plan and financial plan or projection. Parts of the business plan include:

The Mission Statement. The first segment of the business plan is a statement of company mission. This may seem simple because, after all, the immediate plan of any company is to make money. However, most companies have far more specific, or broader, missions than merely making money.

If you have not addressed the subject of the company's direction with your business partner(s), spouse or family, now is the time to do so. Without a definable purpose/goal/mission, businesses of any size tend to wander aimlessly through a maze of daily tasks, moving to and fro with the winds of change in customer demand, financial resources and even the whims of the owner.

From this comes the mission statement. The mission statement in a small start-up company may incorporate some of your strategic goals, but it is primarily aimed at achieving specific objectives this year, or perhaps even next year. This mission statement should be geared more toward operations.

Company History. Also part of the business plan is some background or history of the company or owner. This section should remain basically constant from year to year. Although there is nothing in this section that directly influences the character or the attainability of the plan itself, your history (or the company's history) is a constant reminder of both wise and unwise decisions made in prior years. Having the company's history readily available also has a psychological influence on your employees (who are privy to this document) and others who may read it. Remembering the past tends to bring the present into perspective.

Also, and this point is seldom understood in today's rush-rush business environment — the past usually repeats itself. Keep in mind that one definition of insanity is to do something over and over again, but expecting different results.

NOTABLE SNOWSTORM: March 1941: One of the century's most severe blizzards struck North Dakota and Minnesota, claiming 71 lives.[1]

❄ MISSION STATEMENT: SNOW MANAGEMENT GROUP

"To be recognized as a leader in the landscape and snow removal industries by being an aggressive, efficient and profitable organization. We intend to bring this about by maintaining a quality orientation and by consistently exceeding our customers' expectations.

We recognize the need for continual growth and innovation and will pursue this growth to provide an atmosphere where company personnel can achieve their personal goals through our organization."

Capitalization. The capitalization of a company – the structure of a company's debt and equity (net worth) – is more than just an accounting nicety. It is the foundation upon which to build assets. Typically, a company's capitalization consists of two segments: debt capital and equity capital. Debt capital represents all short- and long-term loans outstanding, plus mortgages. Total debt capital reflects a company's debt and how much cash must be paid out to creditors this year and in future years.

Equity. Equity capital represents ownership in the company and has two segments: 1) amounts invested in the company by owners and 2) earnings from prior years that have been retained in the company's coffers. The relationship between debt capital and equity capital – known as *debt-to-equity ratio* – is a key measure of the solvency of a business and, hence, its ability to meet obligations and to add assets. The higher the debt-to-equity ratio, the less money that is available for operating and growing the company.

Additional Sections. Other important sections of the business plan can include some information on your perceived competition, the services you shall provide, pricing structures, advertising and sales promotional ideas and tactics, customer service programs and collection policies. Another important section would be your management organization. This can be as small as you and your helper. Or, it can be as large and detailed as needed to encompass your office staffing, sales team, field managers, departmental managers and frontline production workers.

Financing Plan. A financing plan is used almost exclusively to project your company's ability to secure and to repay loans. Designed for readers with strong financial backgrounds, financing plans have very little narrative and focus heavily on financial projections. Such projections will cover three to five years and are prepared using the same forecasting techniques as strategic and business plans. Usually this would be prepared by an internal accountant or controller. For the purposes herein, we shall not discuss this at length, as this tool is usually for the very large concern projecting several million dollars in revenues.

Forecasting. Forecasting isn't a plan in the sense of those that are described above. It is a technique for projecting the most probable operating results of a company, in a dollars format, over future periods of time. These periods can be months, quarters or years. Forecasts usu-

SNOW FACT: About 70 percent of winter-storm-related deaths occur in automobiles. The rest are primarily due to heart attacks from overexertions such as shoveling heavy snow or from hypothermia caused by overexposure to the cold.[2]

ally include three basic documents: pro forma balance sheets, pro forma income statements, and cash flow schedules. Forecasts also form the basis for quantifying business valuations when buying and selling a business (or shares in a business), for estate tax purposes, and for many other situations that require a dollar value be placed on a business entity. Again, for purposes of this book, we shall not spend a lot of time on forecasting.

A piece of equipment is a fixed cost and the fuel needed to operate the equipment is a variable cost.

OPERATING BUDGET

An operating budget covers the shortest time period of any plan. The operating budget may be for a week, a month or a year. Operating budgets seldom extend beyond one year. A budget is a formal, numerical expression of how a business (or a division of a business) should operate over the planning time period. Budgets are extremely useful as management tools for planning and controlling segments of a business (like snowplowing).

These segments can include "departments," cost centers, profit centers or functions. The control feature is probably the most useful aspect of budgeting. Actual results for the week, month, quarter and year can be quantifiably compared to expected (or projected) performance, allowing managers/owners to take corrective action in a short time period. While some individuals try to incorporate sales projections into their budgetary process, cost budgets appear to be the most useful in smaller companies. These cost budgets are often called *operating budgets* to differentiate them from more comprehensive planning documents. Since the operating budget is the most common plan used in small companies, we will spend some time discussing this in greater detail.

Fixed vs. Variable. Budget costs are either fixed or variable according to how they change in relation to other areas of the business. All operating costs are fixed or variable. Sometimes they are semi-variable. A fixed cost is one that is independent of other costs, sales or asset changes. In other words, fixed costs will be incurred independently of other transactions affecting the operation. One example is rent. Rent is a cost held constant by a lease.

ACTUAL BUDGET INFO

Below is a page from an actual budget of a snow and ice management company. Such budgets are crucial for effective management.

	Oct-01	Nov-01	Dec-01	Jan-02	Feb-02	Mar-02	Apr-02	May-02	Jun-02	Jul-02	Aug-02	Sep-02	Totals
Snowplowing Service Income	**75,000**	**115,000**	**115,000**	**215,000**	**275,000**	**215,000**	**75,000**	**0**	**0**	**0**	**0**	**0**	**1,085,000**
Gross Payroll	7,183	7,833	7,833	8,267	8,267	8,267	8,267	8,267	8,267	8,267	8,267	8,267	**97,252**
Gross Payroll													**0**
Subcontractor Labor	0	50,000	100,000	125,000	125,000	100,000	10,000	0	0	0	0	0	**510,000**
Office Supplies Expense	445	200	200	200	150	150	100	100	100	100	100	100	**1,945**
Miscellaneous Expense	250	50	50	50	50	50	50	50	50	50	50	50	**800**
Postage Expense	77	50	20	20	20	20	50	50	50	50	50	50	**507**
Printing Expense	14	25	15	10	10	10	15	20	20	20	20	20	**199**
Marketing Expense	250	250	250	250									**1,000**
Travel Expense	845	100					50	100	150	200	200	200	**1,845**
Memberships/Subscrpt/Lic							25	25	25	25	25	25	**150**
Vehicle Expense	155	150	150	150	150	150	150	100	100	100	100	100	**1,555**
Meals & Entertainment	155	25	25	25	25	25	50	100	100	100	100	100	**830**
Insurance (Group Health)	2,800	2,800	2,800	2,800	2,800	2,800	2,800	2,800	2,800	2,800	2,800	2,800	**33,600**
Insurance (Workers Comp)	3,300	3,300	3,300	3,300	3,300	3,300	3,300	3,300	3,300	3,300	3,300	3,300	**39,600**
Insurance (Bldg,Theft,Fire	250	250	250	250	250	250	250	250	250	250	250	250	**3,000**
Insurance (Auto)	500	500	500	500	500	500	500	500	500	500	500	500	**6,000**
Rent Expense	200	200	200	200	200	200	200	200	200	200	200	200	**2,400**
Water & Sewer	100	100	100	100	100	100	100	100	100	100	100	100	**1,200**
Electric	20	20	10	10	10	10	20	20	20	20	20	20	**200**
Gas Heat Expense	20	20	20	20	20	20	20	0	0	0	0	0	**140**
Telephone Expense	378	25	25	25	25	25	25	25	25	25	25	25	**653**
Cable TV Expense	5	5	5	5	5	5	5	0	0	0	0	0	**35**
Vehicle Gasoline	693	350	400	450	450	450	450	400	400	400	400	400	**5,243**
Shop Expense	500	250	250	250	250	250	250	100	100	100	100	100	**2,500**
Computer Hardware/software	200	200	200	200	200	200	200	200	200	200	200	200	**2,400**
Payroll Tax - FICA	550	600	600	633	633	633	633	633	633	633	633	633	**7,447**
P/R - State Unemployment	269	247	247	856	856	247	247	247	247	247	247	247	**4,204**
P/R - Fed Unemployment	19	19	19	66	50	19	19	19	19	19	19	19	**306**
Employee Bonus Program													**0**
Temporary Employee Exp	949	949	949	949	949	949	949	949	949	949	949	949	**11,388**
Semiar & Training Expense									500	500	300	300	**1,600**
Simple IRA Matching Exp													**0**
Interest Expense	50												**50**
Total Expense	**20,177**	**68,518**	**118,418**	**144,586**	**144,270**	**118,630**	**28,725**	**18,555**	**19,105**	**19,155**	**18,955**	**18,955**	**738,049**
Net Profit or Loss	**54,823**	**46,482**	**-3,418**	**70,414**	**130,730**	**96,370**	**46,275**	**-18,555**	**-19,105**	**-19,155**	**-18,955**	**-18,955**	**346,951**

Another example would be comprehensive general liability insurance premiums. These are usually constant for a one-year period. In the absence of a major change, such as the renegotiations of a lease, actual and budgeted fixed costs should be identical month after month.

On the other hand, variable costs are incurred in direct relation to other costs, sales or asset changes. The demand portion of electricity expense goes up or down in direct relation to how much electricity is used each month. Travel expenses are directly related to the number of trips taken in a month or quarter. Fuel expenses are directly related to the number of snow events that occur in the winter time period.

Controllable vs. Non-Controllable Costs. Controllable costs are those that you, as owner, can control to a large extent. That means, as owner, you have the choice of either incurring or not incurring these costs. This would be a decision that often is made without regard for short-term sales volume or production level fluctuations. Some examples of controllable costs would be: the number of people you have on your payroll, the amount of entertaining done, the use of company vehicles, purchase of office supplies or new office equipment. These costs may be fixed or variable over the budget period, but you still have control over whether to incur them or not incur them.

Some costs are beyond the scope of short-term decisions. These would be non-controllable costs. They might include building rent, payroll taxes, property taxes, depreciation and local phone charges.

Generally, non-controllable costs remain fixed for a period of time. However, over two or three years, many can be renegotiated to lower levels or eliminated completely.

CREATING A BUDGET

The goals that budgets seek to meet must be realistic and attainable within the planned time frame. To be effective management tools, operating budgets must not represent pie-in-the-sky dreams. They cannot be contrived targets aimed at motivating employees to reach for the sky. Today's employees do not respond to incentive goals unless they perceive them to be achievable. And it should be noted that one of the main objectives of having operating budgets (if you have employees) is to get your people to improve their work performance.

There is a tendency for owners (who have employees who are privy to the budget numbers) to use budgets as weapons to prevent others from taking certain

SNOW TERM: WINTER WEATHER ADVISORY Issued for accumulations of snow, freezing rain, freezing drizzle, and sleet which will cause significant inconvenience and moderately dangerous conditions.[3]

actions or from doing certain things. For example: "No, we cannot purchase a computer because the budget won't allow it!" when, in fact, you don't need or want new computers.

The author would caution the business owner from doing this. Using budgets in this fashion destroys their effectiveness. It doesn't take the astute employee long to see through such deceptive practices. When this happens, budgets lose their credibility. It will then no longer be an effective management tool.

Budgets should be built from the bottom up and not from the top down. Once you involve your people in the budget process, make sure you give them a timetable that includes time for you to review their budget projections. Ultimately, you (as owner) should approve all budget items, however, this may take time. Building a budget involves good communication and compromise from everyone involved.

❄ RATIOS TO WATCH

PROFITABLILITY	AVERAGE	IDEAL	EXEMPLARY
Profit Margin	42%	45%	60+%
Earnings (EBIT)/Sales	4.6%	10%	22.8%
EBIT/Total Assets	16.4%	20%	40+%
EBIT/Equity	33.7%	40%	60+%
ASSET UTILIZATION			
Sales/Average Receivables	11.37	12	13
Sales/Fixed Assets	7.41	8	8.5
Sales/Total Assets	3.6	4	4.5
LIQUIDITY			
Current Assets/Current Liabilities	1.94	2	2.5
Current Assets/Short-term Debt	6.30	8	10-25
Current Assets/Long-term Debt	2.10	5	10-20
DEBT UTILIZATION			
Total Debt/Equity	1:1	1.5:1	2:1
Long-term Debt/Equity	0.5:1	0.35:1	0.20 or less:1
EBIT/Interest Expense	5:1	10:1	20+:1
OTHER RATIOS			
A/R Days Outstanding	32.9	30	30
A/P Payout Period	41.9	40	33
Quick Ratio	1.5	1.5	1.6
Sales to Working Capital	14.22	15	16

FINANCIAL STATEMENTS

The two basic financial statements applicable to every company, corporation or sole proprietorship is the balance sheet and the income statement.

BALANCE SHEET

Balance sheets are statements in time. They show the values in dollars of the assets and liabilities of a company as of a given date. These values will be different at different times (the day before will be different than the day after). The values of the balance sheet are valid only for that specific point in time.

The "Asset Side." The left side

of the balance sheet shows the dollar value of all the company's assets. Current assets are listed first, followed by fixed, or long-term, assets and then intangible assets. Current assets are:

- Cash
- Receivables
- Inventory
- Prepaid expenses

Fixed, or long-term, assets include:

- Land
- Buildings
- Vehicles
- Equipment
- Long-term investments

Buildings, vehicles, equipment, etc. are depreciated. This means their cost is written off over their useful lives. The amount of depreciation to date is offset against the cost of these assets in a balance sheet. Land and long-term investments cannot be depreciated.

Intangible assets arc comprised of various expenditures for organizational expenses, patents, trademarks, copyrights and so on. In other words, this would be any expenditure that adds value to the company but cannot be touched or held (as with vehicles or equipment).

The sum of current assets, fixed assets and intangible assets represents the total financial assets that a company has at its disposal to generate income or to generate cash.

The balance sheet shows a company's assets, such as this loader, and liabilities, such as the debt (loan) required to purchase it.

The "Liability Side." The right side of the balance sheet shows the dollar value of all obligations or liabilities of the business. These amounts represent amounts owed to all creditors. These creditors include: banks, suppliers, government agencies, etc. The total amount owed to these creditors is called debt.

Current liabilities would include:

- Bank loans
- The portion of any long-term debt that is due within one year
- Accounts payable
- Accrued expenses, which represent expenses incurred but not yet paid for (interest expense, wages, payroll taxes, income and sales taxes, etc.)

Long-term liabilities include:

- Long-term notes
- Mortgages
- Corporate bonds

The sum of current liabilities and long-term liabilities represents the total amounts owed by the company. When total liabilities are subtracted from total as-

FINANCIAL STATEMENTS

Below is a page from the balance sheet of an actual snow and ice management firm.

BILL'S SNOWPLOWING

A Proprietorship

ASSETS	2001	2000
Current Assets		
Cash	$ 98,793	$ 3,156
Accounts receivable	420,467	524,333
Inventory (Salt & Deicing material)	94,504	54,252
Prepaid items	8,029	7,913
Total current assets	621,793	589,654
Equipment and vehicles		
Equipment	896,997	413,235
Vehicles	356,749	350,533
Office furniture	106,917	82,165
Total cost	1,369,663	845,933
Less accumulated depreciation	(620,209)	512,731
Undepreciated cost	749,454	333,202
	$1,371,227	$922,856

BILL'S SNOWPLOWING

A Proprietorship

LIABILITIES AND PROPRIETOR'S CAPITAL	2001	2000
Current Liabilities		
Accounts payable	$ 402,893	$ 157,009
Short-term debt	19,981	45,976
Current maturities of long-term debt	30,597	34,782
Accrued payroll	22,052	11,636
Accrued taxes	53,133	49,547
Other accrued item	3,472	4,069
Total current liabilties	532,128	303,019
Long-term Debt	69,964	49,615
Proprietor's Capital	796,155	570,222
	$1,371,247	$ 922,856

sets, the remainder is called net worth or owner's equity.

THE INCOME STATEMENT

The income statement (at times referred to as the profit and loss statement) shows a series of transactions that have occurred over a period of time – specifically, the period of time between two balance sheet dates. This may be a month, a quarter, a year or any other period of time. However, typically the income statement does not stretch beyond one year. All transactions that have occurred during this period show up in the income statement – sales, costs of materials and labor, operating expenses, taxes and interest paid on loans.

CASH FLOW STATEMENT

There is a third report called a "statement of cash flow." This is not considered a financial statement in an accounting sense. However, it probably tells more about a company's performance than either a balance sheet or an income statement. A properly prepared statement of cash flow shows a company's total cash receipts and cash disbursements during the period of time covered by the income statement.

The relationships among the numbers in these three reports can enable the company owner to analyze the complete financial performance of the firm. These relationships are usually expressed in ratios. Ratios that describe the various relationships among accounts in the balance sheet and income statement are the primary analytic tools of financial analysis. They are used to judge the financial health of a business, to project future earnings and cash flows, and to value a business. Banks use these ratios to determine the viability of a business before lending to it.

The current ratio (current assets divided by current liabilities) is a measure of liquidity. One widely accepted standard is that a current ratio of 2:1 indicates a company has sufficient liquid assets to meet its current obligations. This standard is usually used when applied to larger companies that must follow generally accepted accounting principles. However, small businesses may have hybrid accounting systems that do not conform to these accepted accounting principles. In these cases, some banking institutions may require that the ratio be 3:1 or 4:1 to ensure prudent cash management and adequate liquidity is present.

CASH

Cash is king. We've all heard this statement and it is true. A company that has cash to pay its bills is a well-run company. Current accounting methods permit companies to show a profit on their financial statements even though they don't have enough

cash to meet one week's bills. There is nothing very complicated about cash management. In the simplest terms, cash management is a set of procedures aimed at maximizing the amount of cash available at any given time. We all practice cash management in our daily lives by keeping track of checks we write and deposits we make. Only in this way can we be certain we have enough cash in our personal checking accounts to cover tomorrow's checks. The same principle applies to cash management in our business. We have to make sure the company has sufficient funds in its checking account to pay its bills tomorrow, the next day and the day after. That's all there is to cash management. If we all followed this single, basic principle, there would be no need for this entire chapter.

Unfortunately, most do not practice this principle. Credit is dangled in front of us everywhere. Everyone is encouraged to buy on credit. We borrow to pay credit obligations. And it often gets to the point where the business cannot generate enough cash to meet these obligations. Sometimes it causes good, hard- working people to go out of business. Then they begin practicing good cash management in order to pay the everyday bills.

My advice is to scrutinize every expenditure. Is it necessary? Collect receivables aggressively and have cash on hand.

Some companies take 3 percent of all money that comes in the door and put it away in a separate account.

At each deposit, a check is written to this separate cash account for 3 percent of the total deposit. Though each deposit is small, the money builds up steadily (albeit slowly). It is a great tool to put money aside for unforeseen expenditures that must be made. And the cash balance looks very good on the financial statements as an asset.

Ultimately, a successful contractor is separated from a failed one based on his awareness of his company's financial health. ❄

SOURCES

[1]National Climatic Data Center, National Oceanic and Atmospheric Administration, National Snow & Ice Data Center, University of Colorado, Boulder, Colo., http://nsidc.org. AccuWeather.com, www.accuweather.com.

[2]National Weather Service, Weather Forecast Office, Baltimore/ Washington.

[3]U.S. National Weather Service.

CHAPTER TEN: Working With Subcontractors

IN THIS CHAPTER

- ❄ Advantages and disadvantages of subcontracting snow removal work
- ❄ Legal classification of a subcontractor
- ❄ Subcontractor liability and insurance
- ❄ Finding subcontractors
- ❄ Managing subcontractors
- ❄ Subcontractor agreements

Growth in the snow-plowing business usually depends upon how much equipment and manpower you can muster to deal with the increased business. Additional equipment requires substantial capital investment if you choose to own all the necessary equipment.

One proven method of reducing your capital investment in such growth is to utilize subcontractors. These are others who supply both the manpower and equipment necessary to adequately service your customer base during any particular snow and ice event.

For those who are unfamiliar with how to go about securing, keeping and organizing subcontractors in a growing snow-plowing operation, here are some suggestions and guidelines that might assist you in your quest for growth. Additionally, we shall delve into the legal issues surrounding subcontractors as they pertain the contractor's responsibility and the responsibility of the subcontractor you are working with.

Using subcontractors means you will not need to invest time nor money in an extra piece of equipment.

PLUSES AND MINUSES

There are advantages and disadvantages associated with using subcontractors for snow and ice removal. Obviously, one of the largest advantages is you won't have to invest money in equipment considering the subcontractor shall be utilizing his/her own equipment in the endeavors. This is also one of the tests

Using subcontractors allows a snow professional to serve larger and more numerous clients than might otherwise be possible.

associated with determining if someone is truly a subcontractor instead of being an employee. Using properly classified subcontractors will eliminate the burden of keeping track of and paying the standard employment taxes normally associated with an employee. This alone will allow you greater flexibility in expanding your area of coverage with regard to commercial sites.

Employers must withhold and pay federal payroll taxes for employees. This includes a 7.65 percent Social Security tax and a small federal unemployment tax that comes from the employer.

The employee's own Social Security taxes and federal income taxes must also be withheld from the employee's paychecks by the employer and paid to the IRS. For subcontractors, these taxes do not have to be withheld and forwarded to the appropriate agencies. The subcontractor must handle such taxes and filings on his own, thus lessening the burden on the contractor.

Employers in every state are required to contribute to a state unemployment insurance fund on behalf of employees. These contributions are not necessary when using subcontractors. Workers' compensation insurance coverage is unnecessary for subcontractors.

Sounds great, doesn't it? All these taxes don't need to be paid, all the paperwork associated with having employees is no longer necessary, which will undoubtedly save on accounting and bookkeeping costs.

But it isn't quite as easy as declaring that all of those you hire will now be subcontractors. These subcontractors must be properly classified. You cannot make everyone who works for you a subcontractor just because you want them to be classified as such.

There are several qualification tests that must be satisfied to ensure various governmental agencies do not reclassify your subcontractor as an employee, thus negating all the positives described above.

The IRS may audit you and look very closely at these subcontractors to ensure you are actually treating them as subcontractors and not merely looking to avoid the paperwork that comes with having employees. The IRS wants to see as many workers as possible classified as employees and not as independent contractors. The IRS wants to immediately collect taxes based on payroll withholding. If the IRS audits your business and determines you have misclassified employees as independent contractors, it may impose substantial penalties and interest.

Additionally, the state may also audit your records in order to determine if you have properly classified these plowing subcontractors. They too would like to classify all persons working for you as employees.

WHO IS A SUBCONTRACTOR?

There are a few myths that permeate our society about who is and who is not a true subcontractor. The first one is that "saying a worker is a subcontractor will make it so." Don't get caught in this trap. The mere fact that you call a subcontractor a subcontractor will not by itself convince the IRS or any other state agency that the worker really is one.

Government auditors look at the substance of the relationship between the hiring company and the worker. Mere formalities are not enough. A well-drafted and truthful subcontractor agreement can be helpful in the classification process, but it is not enough to satisfy auditors. Auditors will make sure the parties involved actually behave the way the agreement indicates they should behave. An example of a well-drafted, and tested subcontractor agreement is displayed in this chapter.

The second myth is that people who work for more than one company are considered subcontractors. If you have the right to control the worker on the job, the worker may be classified as your employee even though he works for others at the same time. It's possible for a worker to have more than one employer at the same time. However, if you specifically state that your subcontractor must work for your company and no others, then you will most definitely give any auditor just cause to classify your subcontractor as an employee.

The third myth in this trilogy is the assumption that part-time and short-term workers are subcontractors. Don't think because you're only hiring a person for a couple of weeks the person must be a subcontractor. If you have the specific right to control the person's actions, most government agencies will classify these people as employees.

You can legally use people with their own plow rigs as subcontractors, provided they behave in a fashion that keeps them independent from your snow and ice removal company.

There is a common law test

NOTABLE SNOWSTORM: February 1951: The greatest ice storm on record in the U.S. produced glaze up to 4 inches thick from Texas to Pennsylvania, causing 25 deaths, 500 injuries and more than $100 million damage.[1]

for classifying subcontractors. The common law test is used by most governmental agencies to determine whether workers are employees or subcontractors. This test is used by the IRS, unemployment compensation insurance agencies in most states, workers' compensation insurance agencies in most states, and courts to determine copyright ownership disputes.

THE IRS TEST

The IRS uses an "IRS 20 Factor Test" to explain all the different parameters the agency has historically used to measure control over workers, and thus the determination of whether a person is a subcontractor to your business or an employee.

Recently, the IRS has simplified and liberalized the test. As a result, not all of the factors are as important as they once were in the years past.

Making A Profit Or Loss. Subcontractors can earn a profit or take a loss as a result of the services being performed. Employees are typically paid for their time and labor and have no liability for losses incurred by their employer.

Working On Specific Premises. Subcontractors usually are able to work in many places to perform their services. Employees work where their employers tell them to work, usually on the employer's premises.

Offering Services To The General Public. Subcontractors may (and are encouraged to) offer their services to the general public. Employees offer their services solely to their employers.

Right To Fire. Subcontractors can be terminated for only violating their contractual agreement with the company. Employees typically can be discharged by the employer at any time for a myriad of reasons.

Furnishing Tools And/Or Materials. Subcontractors furnish their own equipment (typically trucks, plows, loaders, deicing equipment, etc.). Employees are typically furnished all the tools and materials necessary to perform their jobs.

Method Of Payment. Subcontractors are generally paid a flat fee for a project. In the snow industry it is customary to pay subcontractors by the hour. And since this is done all over the country and has been done for many years, the practice has stood the test of time.

Where a particular payment method has become general practice, the method of payment factor will not be given great weight. However, it is recommended subcontractors only be paid once per month since employees can be paid (again, usually by a unit of time) each week or every other week.

Also, subcontractors should invoice the company for work performed, and if paid once per month it cannot be erroneously assumed this person is an employee since federal labor laws specifically prohibit paying employees once a month.

Working For More Than One Firm. Subcontractors must have the ability to work for more than one client. This is a common mistake made by snowplowing companies when hiring subcontractors.

Unlike the Method of Payment test, this test carries considerable weight when a governmental agency is determining whether or not a person is truly an independent contractor as compared to an employee. The employee usually is expected to have some loyalty and employers can prevent employees from taking some alternative jobs.

Continuing Relationship. Independent contractors are usually not contracted to work continuously for one company. Usually subcontractors in the snow industry are not hired to work continuously for the same company.

It is not reasonable to expect that a snowplowing subcontractor will be performing snow-related services in summer months. Employees normally have a continuing relationship with the employer.

❄ SNOW TERM: NOR'EASTER

From the Mid-Atlantic Coast to New England, the classic storm is called a Nor'easter. A low pressure area off the Carolina coast strengthens and moves north. Wind-driven waves batter the coast from Virginia to Maine, causing flooding and severe beach erosion. The storm taps the Atlantic's moisture supply and dumps heavy snow over a densely populated region. The snow and wind may combine into blizzard conditions and form deep drifts, paralyzing the region. Ice storms are also a problem. Mountains, such as the Appalachians, act as a barrier to cold air, trapping it in the valleys and adjacent low elevations. Warm air and moisture move over the cold, trapped air. Rain falls from the warm layer onto a cold surface below, becoming ice.[2]

Investment In Equipment Or Facilities. Employees generally have no investment in equipment (or facilities). Conversely speaking, the very nature of the services to be performed requires that the independent snowplowing subcontractor have a significant investment in the equipment being used.

By making such a financial investment, the subcontractor risks losing the investment if the business is not profitable. Owning the equipment also implies that subcontractor has the right to control its use.

Business Or Travel Expenses. Subcontractors typically pay their own expenses (gasoline, insurance, taxes, supplies, maintenance etc.). On the other hand, employees' job-related business and travel expenses are paid by the employer.

Subcontractors decide on their own how to do the work, but snow professionals can and should inform the subcontractors on the customers' expectations.

Right To Quit. A subcontractor is legally obligated (by virtue of the written subcontract agreement) to complete the work he or she agreed to do. Employees may quit the job at any time without incurring any liability to the employer.

Instructions. Employers have the right to give their employees oral or written instructions that the employees must obey about when, where and how they are to work. Subcontractors need not comply with instructions on how to perform their services. Subcontractors decide on their own how to do the work. As brokers of snowplowing services, we can and do give subcontractors some instruction, however, those instructions are usually about the customer's expectations for a completed and acceptable product.

Sequence Of Work. Employees may be required to perform services in the order or sequence set for them by the employer. Our subcontractors generally are not instructed as to the sequence of events required to perform snowplowing on specific sites. However, the customer may have a sequence that is required for adequate and acceptable performance standards. For example, a customer may determine the front of a store should be plowed prior to the rear. This is not considered instruction or sequence on the part of the plowing company, as we are just working under the instructions given to us by our customer.

Training. Independent contractors generally get no training from those that purchase their services. Employees are trained by their employers.

Services Performed Personally. Subcontractors are not generally required to render services personally. They can hire their own employees to do the work. Employees, however, cannot hire others to do their jobs for them.

Hiring Assistants. Subcontractors hire, supervise and pay their own assistants or employees. Employees hire, supervise and pay assistants only at the direction of the employer.

Set Working Hours. Employees ordinarily have set hours of work. Subcontractors generally set their own work hours. In the case of snowplowing subcontractors, it is the customer who requires plowing be completed by a set time so that they can do business. We, as the broker, pass that information on to the subcontractor so we can all meet the expectations of the customer.

Working Full Time. Subcontractors are free to work when and for whom they choose. They also have the right to work for more than one client at a time. Employees, however, are usually required to work full time for that one employer.

Oral Or Written Reports. Subcontractors are generally not required to submit regular reports; they are responsible only for end results. On the other hand, employees may be required to submit regular oral or written reports to the employer regarding the progress of their work.

Integration Into Business. Subcontractors' services are typically not molded into the hiring firm's overall business as one integrated operation. Employees typically provide services that are an integral part of the employer's day-to-day operations.

Worker Benefits. Subcontractors receive no standard workplace benefits. Employees usually receive benefits such as health insurance, sick leave, pensions and/or paid vacation.

Tax Treatment Of The Worker. Subcontractors pay their own taxes and are supplied with a 1099 form attesting to the fact that they have been paid monies without any taxes having been withheld. Employees usually have federal and state payroll taxes withheld by the employer and remitted to the government. Thus employees receive a W-2 at the end of the year attesting to these taxes and wages having been paid.

Intent Of The Hiring Firm And Worker. People who hire subcontractors normally intend to create an independent contractor-hiring firm relationship. People who hire employees normally intend to create the employer-employee relationship.

Custom In The Trade Or Industry. Workers who are normally treated like subcontractors in the industry in which they work are likely to really be independent contractors. This is the norm in our industry, so it is

SUBLIME

Keys To Working With Subcontractors.

By following the basic guidelines and suggestions in this article, you can greatly increase your chances of securing, retaining and organizing subcontractors.

1. HAVE NO FEAR

First and foremost, don't be afraid of working with subcontractors. Working with independent contractors can help you achieve rapid growth without undue investment in equipment, personnel or training.

2. ADVERTISE

Advertising is generally the most accepted method for attracting plowing subcontractors. A couple of lines in the "classified" section of the local paper will typically elicit inquiries. This can be as simple as *"Snowplowers, with vehicle, needed. Routes throughout* (your market) *area. Call: 555-1234 for information."*

Those individuals who respond will have some specific concerns and questions, and you should be prepared to address these inquiries. These questions often will include: "How am I getting paid?; When do you pay?; How do I get told when to plow?; Where will I be plowing?"

3. ASK FOR REFERRALS

Another effective method to attract additional subcontractors is establishing a referral program. Such a program allows existing subcontractors to "refer" your company to other potential subcontractors.

Snow subcontractors generally talk with one another on citizens' band (CB) radios during snow events, and they get to know the good contractors. If you pay a referral fee to your existing subcontractors and if you have treated your subs fairly (meaning you paid them on time and in full, as promised), you'll soon find existing subcontractors recruiting others to come work with you.

One suggestion would be to pay $200 to an existing subcontractor who refers another subcontractor to work with your company. You could pay half the referral fee up front and the balance at the end of the season if the new sub has worked well over the entire season.

4. DEVELOP A STARTER KIT

Develop an information packet that can be mailed to the potential subcontractor. This packet should include your policy on insurance, a formal subcontractor agreement that lays out the sub's responsibility as well as your own, the pay rate or pay scale, and an explanation of your company policy about how you generally deal with subcontractors. This packet of information can be mailed to the prospective subcontractor after he/she first contacts you.

5. MAKE PAYMENT A PRIORITY

Subcontractors should be paid on time and in full, every time. Failing to pay a subcontractor on time only leads to bad publicity over the CB airwaves. Subcontractors know who's paying on time and who isn't. The word gets around, both good and bad. Some companies have subcontractors waiting in line to work for them, simply because they have earned the reputation as a company that pays on time and in full. This is a big deal to subcontractors, so don't underestimate the power of paying in full and on time.

Subcontractors should also be paid on a scale in accordance with the equipment and experience they possess. Start with a base rate that is fair, then pay more money for a larger or more efficient plow (i.e., the 'V' blade, snow wings or capture blade), a bit more money if they carry a cellular telephone, and more money still if this is the subcontractor's second or third year with your company.

Further, more money can be paid for subcontractors who have trucks with an automatic transmission instead of a manual transmission. No matter how much the subcontractors plead with you, they can't back up as fast with a standard transmission as they can with an automatic transmission.

6. SUBCONTRACTORS ARE NOT EMPLOYEES

While you want subcontractors to "feel at home" and be happy to work for you, be very careful about how you treat subcontractors while they are actually plowing snow for you. You do not want to give anyone, including your own employees, the impression they are your employees.

7. MAKE SURE YOU ARE LEGAL

The IRS and many states have regulations governing how relationships with subcontractors must be handled. Make sure you know these guidelines and follow them.

8. ADVISE EXPECTATIONS

Though you cannot instruct subcontractors like a traditional employee, you should make sure the subcontractor is thoroughly informed on the customer's snow removal expectations. Usually your customer does not recognize the difference between you and a subcontractor. As a result, the actions of your subcontractors will affect the reputation of your business. Make sure the subcontractors you use measure up to your customer's expectations.

9. FREEDOM

Don't restrict subcontractors from plowing for themselves or other contractors. If you treat them right and pay them as promised, they won't go elsewhere anyway.

10. AVOID LOGOS

Never put your logo or sign on a subcontractor's truck or on a piece of subcontracted equipment. Remember, subcontractors are not your employees, but rather are independent contractors providing you a service for a set fee.

11. AT YOUR OWN EXPENSE

Do not pay any of the subcontractor's expenses other than the agreed upon rate of pay.

12. PROPER COVERAGE

Require proper insurance coverage from any and all subcontractors you work with.

13. PROPER DOCUMENTATION

Never pay any sub until he has filled out an IRS Form W-9 for your files. Always issue a proper IRS Form 1099 to the subcontractor at year end and make certain the subcontractor knows this income will be reported to the government, as is required by law. ❄

likely they are actually independent contractors. This is the same for employees. Workers who are normally treated like employees in a particular trade or industry, are likely to actually be employees.

KNOW YOUR ABCS

About half the states in the United States use a special statutory test, also called the ABC test, to determine if workers are subcontractors or employees for purposes of unemployment compensation.

This test focuses on just a few factors. These factors are: whether the hiring firm controls the workers on the job; whether the worker is operating an independent business; and where the work is performed. This test is much simpler than the IRS test, but can be the hardest to fulfill its requirements.

There are numerous other factors involved when setting up subcontractors to work with you in your endeavors to grow your business.

This can be a subject of an entire book in itself. If you would like to know considerably more on the subject, I recommend you obtain *Hiring Independent Contractors* by attorney Stephen Fishman. Some practical advice and information to keep in mind when dealing with snowplowing subcontractors is noted in the text below.

INSURANCE AND SUBCONTRACTORS

Subcontractors in the plowing business should have insurance. It is in your best interest to have your subcontractors carry insurance, so they are responsible for any damages they create and for the portion of any liability that is assigned to your company in the event such an issue arises. Some states require vehicles driven on roads have a minimum amount of insurance coverage. In those states where it is not mandatory, you should insist on it for your own (and others') protection from catastrophic events that might take place.

If the subcontractor has employees, you should insist those employees carry workers' compensation coverage at state mandatory minimums. Keep in mind that if a subcontractor's employees are hurt in some fashion on the job and if the sub is not carrying proper insurance coverage, you could be held liable for covering their injuries.

In addition to the above-referenced insurance coverage, the subcontractor should carry comprehensive general liability insurance. This insurance policy covers accidents that may happen after your sub has left a site and has completed the snowplowing operations. While you (the plowing contractor/broker of record) can take steps to limit the exposure of such liability in your contract language, you cannot totally divorce your

SNOW FACT: Snow flakes are actually six-sided ice crystals. The temperature at which the crystal forms, and to less extent the humidity of the air, determine the basic shape. The many things that happen to snow crystals as they fall, such as collisions and partial melting, create shapes. Snow crystals are typically: thin plates at 32 to 25 degrees F; needle-like at 25 to 21 degrees F; hollow columns at 21 to 14 degrees F; segmented plates at 14 to 10 degrees F; and dendrites at 10 to 3 degrees F.[3]

Independent Contractors Agreement
Snow Removal Services

TERM: This agreement is made between ______________________________

______________________________ DBA ______________________________ (called "Contractor").
(Individual's Name) (Company Name)

It will be effective commencing November 1, __________ and continuing until May 15, __________.

OBJECTIVE: The Company is in the business of providing snowplowing and removal, and similar services. Over the years it has developed a large number of accounts throughout Erie, PA and the surrounding suburban area and is continuously adding additional accounts for these services. These accounts are serviced primarily by The Company brokering out the work to independent contractors who provide their own equipment and perform the actual work. The intent of this agreement is to establish a working relationship between The Company and Contractor for this purpose. In order to establish and effectuate this relationship, the parties agree as follows:

1. **CONTRACTOR SERVICES.** Contractor will be engaged by The Company as an independent contractor and agrees to provide services for The Company's accounts as directed by The Company during the period stated above. Contractor has the following equipment which will be provided, along with a qualified operator, at all times when requested by The Company:

EQUIPMENT

Year	Make	Model No., Transmission Color/Description	Size of Blade/Bucket	Rate per hour

2. **PAYMENT.** The Company will pay the Contractor the rates shown above for the Contractor's services and/or equipment. Payment will be made about 10 days after receipt of invoice from the Contractor for services rendered in the prior month. It is recommended that invoices be submitted on (or about) the 1st day of each month. The Company requests only one invoice per month and under no circumstances will checks be issued more than once per month. No checks will be issued between the 1st of each month and the 10th of each month. In the event there are any damages done by the Contractor to the property of The Company's customers, or poor workmanship that has caused The Company to have to credit the customer's account, these damages will be paid for by the Contractor and shall be deducted from any amounts due the Contractor.

3. **TRAVEL TIME.** Contractor will be paid for travel time between the various jobs that have been assigned to him. Agreed rate will begin at time Contractor leaves his home (in plowing vehicle and ready to plow) and will end at completion of the last job. Special arrangements will be made if Contractor lives outside Erie as travel time to first job will not be paid for long distances. No travel time will be paid from the last job site on the assigned route. The Company does not pay for "breaktime" or "downtime". The Company does not pay for time required for fueling the Contractor's vehicle. If vehicle must be fueled prior to starting the assigned route, the start time will begin when Contractor leaves fueling point.

4. **CONTRACTOR'S EXPENSES.** Contractor will provide all fuel for his equipment. All maintenance will be at Contractor's expense and his equipment is to be properly maintained to avoid breakdown during an urgent snow removal period. Contractor will not be paid when his equipment is broken down or inoperable.

5. **SUBCONTRACTOR AND INSURANCE.** Contractor is engaged as an independent contractor and will not be considered an employee of The Company. Contractor will provide his own equipment as noted above, as well as any other tools or supplies which are necessary in order to provide the services which Contractor is engaged to provide. He must provide transportation for himself and his employees. He must have a telephone where he can be reached by The Company. He must provide any office or administrative services which he requires apart from The Company property and without cost to The Company. He must provide his own Worker's Compensation insurance for himself and for his employees. Contractor must provide auto liability insurance with recommended limits not less than $300,000. He will provide The Company with appropriate certificates of insurance. He will be responsible for payments to his employees for their work and for any required withholding, and pay overtime according to law. If, for some reason, additional insurance premiums and/or taxes are levied against The Company because of the services provided by Contractor, then those costs will be paid to The Company by Contractor. The Company may withhold such costs from any amounts that it has due to the Contractor from The Company. If there are not sufficient amounts owing Contractor by The Company to cover such costs at the time they become due by The Company, then Contractor shall repay them to The Company within 30 days from his receipt of notice thereof from The Company. If payment is not made, The Company may proceed to collect from the Contractor by legal process and in such case the Contractor shall also be responsible for The Company's reasonable costs and attorney's fees. In any event, failure to provide a certificate of insurance will result in 25% of amounts due to Contractor being retained by The Company to cover additional insurance premiums that may be levied against the Contractor by his insurance company in order to adequately insure that apparent uninsured Contractor. This retainage will be paid to the Contractor within 30 days of receipt of a valid certificate of insurance, and will be paid in the following manner; 15% of retainage due to be paid within 30 days of receipt of actual "Certificate of Insurance", 10% retainage to be paid if Comprehensive General Liability coverage in the amount of $100,000 is shown on the certificate.

F:\Netserver\wanda\snow-contracts\snow-sub-agreement-AC-page1.doc Updated: 5/7/2002

Example of a subcontractor agreement.

6. SUPERVISION. Contractor will be responsible for the work assigned to him and for the conduct of himself and his employees and will provide adequate supervision for them. He will assure that his work and the work of his employees is provided in a good and workmanlike fashion and in accordance with the standards established by The Company. Within these limits, Contractor may perform the job as he sees fit. He agrees to coordinate with The Company and that The Company has the right to oversee the quality and timeliness of the work. However, The Company will have no obligation to provide supervision or control of Contractor's manner of performing the work or of his employees.

7. STARTING TIME. Time is of the essence to The Company's customers. Therefore, the equipment shall be "rolling" within 30 minutes of the time that The Company gives notice to the Contractor. The Contractor shall keep aware of weather conditions and when there is threatening weather, shall be ready to roll on a moment's notice. In the event of the possibility of snowfall the Contractor agrees to keep The Company informed of whereabouts and availability if Contractor will not be at his regularly designated phone number.

8. MINIMUM USAGE. The Company will not estimate the hours of usage for the above equipment nor guarantee a minimum usage, due to the uncertainty of winter weather patterns.

9. NEW ACCOUNTS. It is understood that The Company has a large number of accounts for which it has done work over the past years and that it is continuously engaged in advertising and sales promotions in an effort to acquire new accounts. Therefore, all accounts for which work is done are accounts of The Company. Contractor may solicit new accounts for The Company. Such new accounts will be accounts of The Company even though they were procured for The Company by Contractor. Contractor will receive a commission for all new accounts secured by him, which commission will be equal to 5% of The Company's gross billings to that account for the first year. The commission will be paid on May 15th, providing that The Company has been paid in full by the account.

10. NON-COMPETE. Because The Company has expended many years and large amounts of money developing and maintaining the accounts which it has, and because Contractor will be placed in contact with many of those accounts in performing the services under this Agreement, Contractor agrees that he will not compete with The Company as an employee, Sub-Contractor, Contractor or in any other capacity, by providing the services which are the subject of this Agreement for any of The Company's customers for a period of two (2) years after the termination, for any reason whatsoever, of his engagement with The Company as an independent contractor. In addition, Contractor will not disclose to any other person, or use on his own behalf, any customer names or other confidential information gained from The Company. The parties agree that in the event of Contractor's violation of this agreement not to compete, The Company will be entitled to both temporary and permanent injunctions and, at its option, liquidate damages in an amount equal to the total gross amount which Contractor earns from competing jobs during the two (2) year period after termination with The Company, plus The Company's costs of enforcing this Agreement including its Court costs and attorney fees.

11. BILLING AND ADVERTISING. The Company will be responsible for billing customers for the work done by Contractor hereunder and for collecting from those customers. It will continue its advertising and sales promotion activities in order to obtain new accounts and maintain existing accounts.

COMPANY:

By: ______________________________

Title: ______________________________

Date: ______________________________

12. RADIOS. The Company may, at is option, provide a two-way radio for use in Contractor's equipment, if his route requires one, without charge, if Contractor will agree to pay the installation fees. These radios will remain the property of The Company. Contractor will be responsible for loss or damage of radios or any other leased equipment. Contractor will be required to add this piece of equipment to his automobile insurance policy and have the Company named as "Loss Payee" in case of loss or damage of radio(s).

13. TAXES. The Company will not be responsible for any withholding of income taxes, Social Security, or the like from amounts paid to Contractor under this Agreement. It will not be providing Worker's Compensation Insurance for Contractor. It will not provide any company benefits of any kind for Contractor. Its obligation to Contractor will be solely those set forth in this Agreement. The Company will send IRS form 1099 to Contractor at his last known address, after the end of the calendar year as provided by law.

14. TERMINATION. It is the intention of the parties that this Agreement, and the relationship between them, will continue until terminated by the parties. Either of the parties has right to terminate the Agreement at any time by giving written notice to the other. However, when The Company obtains a job or route and offers the work on that job to Contractor, then if Contractor accepts the work, he will not terminate the Agreement, except for cause, until that job or route is completed. The Company may terminate the Agreement immediately at any time for cause.

15. CHARGING AGAINST COMPANY. Contractor will not use the name of The Company or attempt to use The Company's credit, for the purpose of buying or charging fuel, parts, tires or anything else that would inure to Contractor's benefit. If anything is charged to The Company without authorization, Contractor can expect The Company to take whatever legal action is necessary to seek reimbursement and will be responsible for all expenses incurred by The Company. Also this would be cause for termination of this Agreement by The Company.

16. DAMAGES. Contractor understands that he will be fully responsible for any damages to the property of any of The Company's customers which is done by Contractor or any of his employees. The Company may withhold funds to repair such damage from any amounts owed by The Company to the Contractor. If the amount of the damage exceeds any amount owed by The Company to the Contractor at the time the amount of the damage is ascertained, then the Contractor shall pay over to The Company the difference within 30 days thereafter. If payment is not made, The Company may pursue collection by legal means and the Contractor shall then also be responsible for The Company's costs and attorney fees.

17. MISCELLANEOUS. This Agreement is made in the Commonwealth of Pennsylvania and will be construed according to Pennsylvania Law. If any part of the Agreement be found invalid, it will not invalidate the remainder of the Agreement. The Agreement may be amended only by a writing signed by both parties.

SUBCONTRACTOR:

____
Company Name

____
Street Address

____
City State Zip

By: ______________________________

Title: ______________________________

Date: ______________________________

F:\Netserver\wanda\snow-contracts\snow-sub-agreement-AC-page2.doc

Updated: 5/7/2002

A couple of lines in the classifieds will usually produce calls from a few interested subcontractors.

company (or your subs) from the possibility of someone filing suit against you.

Most insurance-savvy customers will insist on being named "additional insured" on your policy. This is normal and to be expected. The plowing contractor should insist on this same "additional insured" status on the subcontractor's insurance policy. This is normal, and should not cause a problem with the subcontractor's insurance carrier or agent.

FINDING SUBCONTRACTORS

Advertising for subcontractors is generally the most accepted method for attracting plowing subcontractors. A couple of lines in the "classified" section of the local paper will certainly get some inquiries. This can be as simple as "*Snowplowers, with vehicle, needed. Routes throughout* (your market*) area. Call: 555-1234 for information.*" Those individuals who respond will have some specific concerns and questions and you will need to plan for how you will address these questions. The questions will undoubtedly include:

- How will I be paid?
- When do you pay?
- How do I get told when to plow?
- Where will I be plowing?

It is a good idea to have an information packet put together (in advance). The packet can be mailed to the potential subcontractor and will go a long way toward eliminating any misconceptions later in the winter. This packet of information can be mailed to the prospective subcontractor after he/she calls. This packet should include:

- Your policy on insurance
- A formal subcontractor agreement that lays out the sub's responsibility as well as your own
- The pay rate or pay scale
- An explanation of your company policy about how you generally deal with subcontractors.

Subcontractor Referral. Another method to attract additional subcontractors is to have a referral program in place that allows existing subcontractors to "refer" your company to other potential subs.

Subcontractors all talk on CB radios during snow events, and they get to know who the good contractors are by talking with each other. If you pay a referral fee to existing subs, and if you have treated your subs fairly (meaning you paid them on time and in full, as promised), then the referral fee can serve as a good incentive for existing subs to refer new subs to you.

One suggestion would be to pay $200 to the existing subcontractor who refers another to come work with you. The new sub would have to work the entire season. You could pay half the referral fee at the beginning or middle of the the winter season and the balance at the end.

MANAGING SUBCONTRACTORS

Using subcontractors can be a source of irritation at times, and one needs to be cognizant of the pitfalls associated with working with them. One also needs to be prepared that a percentage of the available subs will not come out when called. The reasons for these no-shows can include: sickness, hangover, broken truck, no babysitter and the phone being turned off (inadvertently, of course).

Subcontractors often network with one another via CB radio, so a referral program can help you find additional help.

However, subcontractors normally take very good care of their equipment. Subs don't normally ram curbs with plow trucks and then say "oops." And when a subcontractor's equipment breaks down, he normally works very hard to get back on the road immediately instead of calling on the radio (or phone) and saying,

Most important is the manner and timing in which you pay subcontractors. If you do not pay subcontractors fully and on time, every time, word will spread quickly.

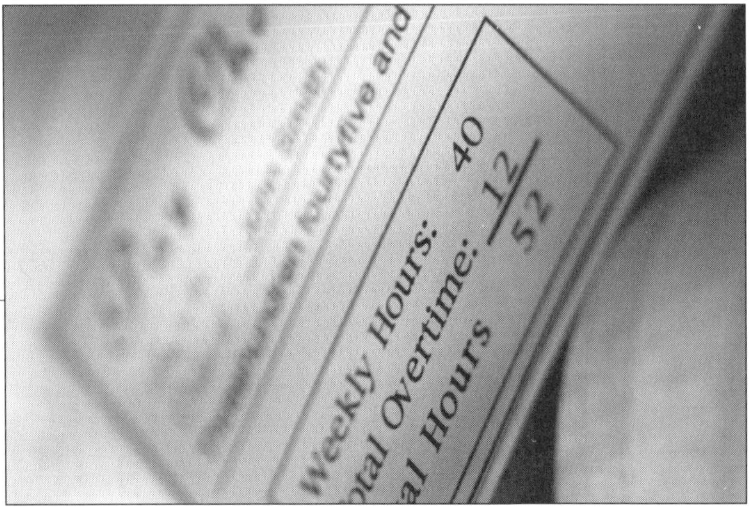

"My truck's broken." Subcontractors will often carry spare parts and tools to effect repairs immediately so they can get back to work, earning money.

When subs are done for the night (or daytime, as the case may be), they should report hours the same day. In this manner, if there are any discrepancies in recorded hours, they can be immediately addressed. If not, you can end up arguing later when no one actually remembers what went on during that particular snow event.

Sometimes this means calling them at home and waking them up even as they are trying to get some sleep. It is better to address this particular issue right away, rather than waiting.

Show Me The Money. The manner in which a subcontractor is paid is a very big issue. It should go without saying that your subcontractors are the lifeblood of your business, and you need them to survive. Accordingly, they should be paid on time and in full, every time. Stretching out a subcontractor only leads to bad publicity over the CB airwaves. The word gets around. Some companies have subcontractors waiting in line to come work for them, simply because they pay the subs on time and in full. It is a very big deal to subcontractors.

Subcontractors should be paid for the equipment they have available for your use. Consider a "differential system" depending upon what type of equipment the subcontractor owns. Start with a base rate that is fair for your market area. Then pay more money for a larger or more efficient plow (i.e., the 'V' blade, snow wings or capture blade), and even more money for high-capacity equipment (i.e., push plow, front-end loader, dump truck, etc.). Pay a bit more money if the subcontractor carries a cellular telephone. (You should then call the cell phone at least once during each snow event so you know he actually has the phone with him.) Pay the subcontractor more money for his second and/or third year with your company and more money still for having a truck with an automatic transmission instead of a manual transmission, and so on.

SUBCONTRACTOR AGREEMENTS

The use of written contracts lets all parties know what is expected of them. Additionally, such a contractual arrangement satisfies one of the IRS "tests" for determining if a person is a subcontractor or an employee.

This contract can be as detailed as necessary. It should specifically state the subcontractor is not an employee. This satisfies the "intent of the parties" portion of the IRS test procedure and policy.

Additionally, it should detail what equipment the subcontractor will use while plowing the sites that will eventually be assigned to them. This detail should include the type of equipment, type of plow, rates charged for specific types of equipment, and the subcontractor's company name.

Additionally, the contractor should detail what is expected of the subcontractor in terms of availability, response time, hours available for work, any guarantees on how much work there might be available, payment terms, what documentation is necessary in order to be paid, the fact that no reimbursement for expenses incurred will be forthcoming, insurance requirements, responsibilities with regard to the customers' expectations, and the like.

The use of subcontractors is a tool for growth. And, like any valuable tool, it needs to be used effectively.

Subcontracting is not the only way to grow your snowplowing business, but it may be one of the easiest methods for growth without substantial capital investment on your part. Treat subcontractors right, and they will be loyal and productive members of "your team." ❄

SOURCES/FURTHER READING

[1]National Weather Service.

[2]National Snow & Ice Data Center, University of Colorado, Boulder, Colo., http://nsidc.org. AccuWeather.com, www.accuweather.com, National Climatic Data Center, National Oceanic and Atmospheric Administration.

Hiring Independent Contractors by Stephen Fishman.

PART TWO

IN THE FIELD

CHAPTER ELEVEN: Snow Removal Equipment

IN THIS CHAPTER

- ❄ Finding a plow truck
- ❄ Selecting a plow
- ❄ Equipment maintenance
- ❄ Using heavy equipment

Selecting, maintaining and operating snow removal equipment can be one of the most important steps a snow professional undertakes. Without solid employees and reliable equipment, no customers can be served. Equipment for snow removal is highly varied, ranging from the tried and true snowplow mounted on a pickup truck to the use of heavy equipment, such as front-end loaders and large dump truck units.

All major truck manufacturers have their strong and weak points when it comes to plowing snow. Pictured from left to right is a Chevy, Dodge, Chevy and Ford truck.

THE PLOW TRUCK

When it comes to putting plows on trucks, there are as many different opinions as there are dealers of plowing equipment. It can be confusing. Most large contractors will agree (for the most part) the equipment on the market today will do the job, no matter which manufacturer's product you select.

Some confusion in the marketplace can be traced to the addition of the driver's side air bag in trucks. It is a federal law that all trucks weighing less than 8,600 lbs. gross vehicle weight must meet the same safety standards as cars. This makes automotive engineers very conservative. Also, more than 90 percent of pickups and sport utilities sold today are sold to car buyers who want a "new car ride" in their trucks, as well as all the options available in cars. Because of this, the stripped-down work truck is a thing of the past.

Trucks are being designed with a softer ride and are now made closer to the ground so people can get into them easier. Much of this has to do with ½-ton trucks and sport utilities, so truck manufacturers now consider sport utilities, ½-ton trucks and smaller to be personal use vehicles, not commercial truck vehicles.

Below is some feedback from various manufacturers about their recommendations regarding mounting plows on their trucks:

- **Chevrolet/GMC**: Chevy approved 10 models (only two are ½-ton trucks) and all of them must have the snowplow prep package. Chevy GMC wants no plows on any sport utility vehicles, no plows on extended cab trucks, crew cab trucks and Suburbans.

- **Ford**: When the new Ford F-150LD first came out, Ford warned against mounting a plow on the truck, but has backed off somewhat. To use/mount a plow, the truck must have a 3,900 lb. front axle, the plow cannot exceed 500 lbs. and the truck must have the Heavy-Duty Service Package. On the F-250LD, Ford only approves a plow for a 3,950 lb. front axle. It does not approve of plows on any sport utility vehicle.

- **Dodge**: Only trucks with gas engines are approved for plow mounting. And these must have AHD plow packages. No ½-ton trucks approved.

- **Toyota**: No plows are approved on any trucks.

Heavy Duty. As you can easily see, the truck manufacturers are saying that if you want a plow, get a ¾-ton truck or larger. From a practical point of view, this is good advice. The primary issue of concern is the amount of weight placed on the front end of the truck. If a pickup is designed to carry three passengers, then three 150 lb. people must be figured into the equation to see if the front axle is overloaded. And overloading is a problem not to be taken lightly because it can affect wear and federal safety standards in braking.

If you have a four-passenger sport utility vehicle (or extended cab) or even a six-passenger Suburban (or crew cab), the problem is worse. But what if you plow alone and thus are not carrying the weight of three or five passengers? This is not considered in the equations. There is another point they have not added to the equation. That is rear ballast. The more weight (within limits) that is added to the rear axle, subtracts weight added to the front axle by adding a plow.

SNOW TERM: WINTER STORM OUTLOOK
Issued prior to a Winter Storm Watch. The outlook is given when forecasters believe winter storm conditions are possible and is usually issued 48 to 60 hours in advance of a winter storm.[1]

Unapproved Truck? The question then arises, "What can happen if you put a plow on an un-approved truck?" The manufacturer's warranty on the front-end suspension will be voided. The brakes may also prematurely wear, causing obvious safety and liability concerns.

There are dealers who will install plows on almost any vehicle out there in the marketplace, but this causes problems if an accident occurs. Using a plow on an unapproved truck can create serious liability questions in the event of an accident. If a fatality occurs due to this overloading, the emotional devastation can be a significant issue.

2WD vs. 4WD. Another concern is whether or not a plowing vehicle needs to be four-wheel drive (4WD) or two-wheel drive (2WD). Many Departments of Transportation (DOT) around the country use two-wheel drive trucks, often with V-box salt/sanders mounted in the beds of the trucks. For DOT work, the 2WD truck is fine, as the vehicle is always moving and inertia will assist in keeping the vehicles from losing traction.

Private snow removal contractors encounter mainly parking lots and driveways, which require frequent stops and starts and changes in direction. In these instances, it is almost always beneficial to use 4WD vehicles instead of 2WD vehicles for plowing operations. Some additional maintenance is involved given the nature of having 4WD hubs, a front driveshaft and from the inherent problems that arise from plowing snow.

Vehicles with 4WD move through deep snow with little additional effort and will allow the stacking of snow much easier than with a 2WD vehicle.

Electrical Systems. Electrical systems on plow trucks need to be carefully evaluated, as snowplows draw considerably more power than standard truck equipment. Some plow manufacturers recommend alternator units that put out a minimum of 74 amps. Even this may not be enough in some situations when you add lights to the plow (or safety bubble) and other accessories. Many plowing contractors add a second battery and upgrade the alternator significantly when adding a plow to the vehicle.

Transmissions. Transmission selection can lead to spirited discussions when snow professionals get in the same room. While some prefer a standard transmission, most will agree that an automatic transmission is more efficient during plowing operations. Automatic transmissions allow for faster speeds in reverse.

Further, most automatic transmissions will not require you come to a complete stop prior to shifting into reverse. Standard transmissions will grind incessantly if you attempt to shift into reverse prior to stopping the vehicle.

Additionally, plowing with the automatic transmission in "low" is recommended. This is because if the shifter is in "drive" or "3," the transmission will automatically try to shift up or down depending upon RPM and load. It is possible the transmission will go through all the gears (sometimes several times) during one plow run down a parking lot depending upon the weight of the snow and the resistance it creates. This constant shifting is hard on the transmission and can greatly shorten its life. Plowing in "low" only will not allow this constant shifting and pounding of the linkage. This is not to be confused with the "low" range on the 4WD shifter. "Low Lock" on the 4WD shifter is for use in very, very deep snows. Most of the plowing done will be in "4WD High" with the transmission shifter in "low."

If the truck is outfitted with "Overdrive," the overdrive must be disconnected prior to beginning plowing operations. Plowing with the overdrive engaged will quickly burn out the transmission. This is an expensive lesson that can be avoided by paying attention when starting up the vehicle.

Even if you are careful, the automatic transmission will heat up during plowing operations. If the truck does not already have one, installing a transmission cooler is a wise investment. Keeping the transmission cool should be a top priority when considering plowing in a truck.

Leafs vs. Springs. According to Chuck Smith, owner of the Internet site www.snowplowing-contractors.com and considered one of the foremost authorities on plow trucks and related equipment, mounting a plow on a vehicle with torsion bars is not recommended.

He also recommends avoiding putting plows on vehicles with coil springs in the front end. Leaf springs are much better, as they are designed to carry

SNOW FACT: Most heavy snowfalls typically occur with relatively warm air temperatures near the ground, typically 15 degrees F or warmer. Warmer air can hold more water vapor.[2]

heavy loads. Further, leaf springs can be easily replaced and easily repaired because nearly every spring shop can make new ones or replace broken leaves. You can also add a leaf to compensate for the dip the front end takes with the additional weight of an attached plow.

However, Mr. Smith reminds us that adding leaves doesn't increase the front axle's carrying capacity. The extra leaf compensates, but doesn't strengthen the axle in any way. The axle itself can still only carry what it was designed to carry.

SELECTING A PLOW

When considering the purchase of a plow for your truck, you have some decisions to make. The amount of money you have available will be a mitigating factor in the decision-making process. But, like most things, you often get what you pay for. Plows are manufactured for specific uses. Light-duty plows are inexpensive, and while usually ad-

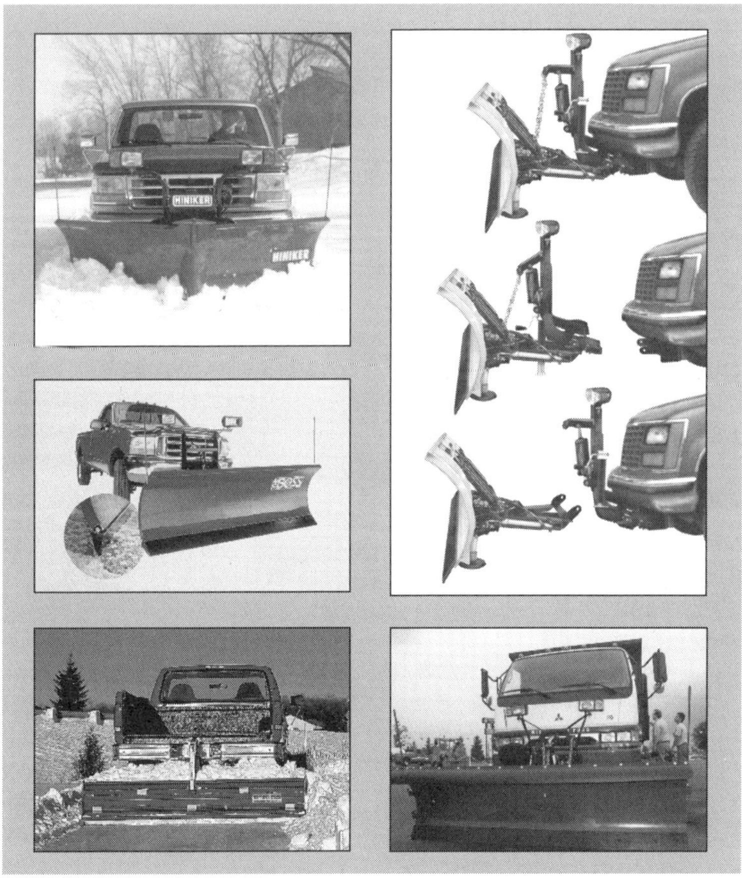

A variety of snowplow types and features are available including (clockwise from left to right): V-plow, specialized mounting systems, polyurethane edges, rear-mounted "pull" plows and trip-edges.

KEEP ON TRUCKIN'

Tips on choosing the right vehicle for snowplowing.

Ask a group of snow contractors about their favorite vehicle for plowing and you're not likely to get the same answer twice. Indeed, the types and makes of vehicles used for snowplowing can be as individual as snow contractors themselves.

"I really don't think there is a 'right' vehicle that will satisfy everyone," said Alan Nadeau, Nadeau Lawn and Landscape, Milton, Vt. It's impossible to find a one-size-fits-all vehicle that works for each contractor because finding the right vehicle depends on the individual operation.

"Get the vehicle that suits your needs best. Just because it works for someone else, doesn't mean it will work for you," said Charles Smith, AC Grounds Maintenance, West Orange, N.J.

Obviously, a contractor serving a large commercial customer would need access to a variety of heavy equipment from loaders to skid steers to dump trucks; while a contractor serving the residential market might do well with just a modestly equipped pickup truck.

ITEMS TO CONSIDER. While vehicular options abound, there are several basic issues to consider when selecting a pickup truck for snow removal operations.

Fuel Type: There are two basic options to choose from: gasoline- or diesel-powered engines. Many contractors reported they prefer to purchase diesel-powered trucks because they typically offer increased power, added reliability/serviceability and, most importantly, increased fuel economy.

Other contractors aren't convinced diesel is the better option. "I'm not real big on diesel power. I like the fuel economy, but the weight of the diesel is just that much more on the front axle," Nadeau said. "With the weight of the plow, you're already heavy up front in most applications, and you really notice it if you get into soft ground."

Two- or Four-Wheel Drive: There are some contractors who plow with two-wheel drive trucks, but the vast majority advise using four-wheel drive. In fact, with the treacherous conditions snow contractors face during a heavy storm, four-wheel drive is, in general, a requirement.

Pat McTavish, MacKenzie & McTavish Snow Removal, Inc.,

Silver Spring, Md., uses a four-wheel-drive truck but switches to two-wheel-drive when he uses the truck for personal travel. For contractors who use their trucks similarly, McTavish recommends getting manual locking hubs, rather than the automatic versions. In McTavish's model of truck, the automatic hubs are made of Teflon and can be easily stripped. Before he finally converted to manual hubs, McTavish had replaced the automatic hubs four times.

Transmission: More than any other "under the hood" component, transmissions take the biggest pounding from snow removal operations. Tom Peters, Northwest Snow & Ice, Schaumburg, Ill., has replaced his transmission four times. "In a lot of these new trucks, people are finding that the transmission can't handle the torque of the diesel motors," Peters said.

"The worst enemy of an automatic transmission is heat," according to Smith. "Anything you can do to help your transmission run cooler will extend its life."

Smith suggests the use of an auxiliary transmission cooler and allowing the truck to idle a few minutes after plowing before shutting it off. Shutting the engine off immediately after operation stops transmission fluid from circulating, and thus, prevents cooling the transmission. Preventive maintenance, such as draining and changing transmission fluid, should also be completed. Remember that because you are putting more than average strain on your vehicle, you should change the fluid more frequently than what is recommended in the owner's manual for "average" use.

Most experienced contractors agree that automatic transmissions are the preferred choice because they are easier to operate, given the frequent forward and backward changes required for snowplowing.

Suspension: Selecting a truck with "Heavy Duty" specifications is preferred for snow removal operations. While the ride may not be as smooth, the demands placed on the vehicle from snow removal usually calls for a sturdier construction. "People don't buy heavy duty trucks to ride like cars. I don't want it to ride like a car, I'm buying it to get the job done," Peters said.

"Plowing definitely qualifies as severe service, so I feel it's important to spec heavy duty, especially if you plan on carrying a salt/sand spreader, as well as the plow," said Rob Nagle, a snow contractor in Orillia, Ontario, Canada.

In general, a ¾-ton or heavier vehicle should be used, and leaf springs are generally preferred. "Leaf springs are best for a plow truck," reported Smith. "Coil springs are good, and I wouldn't mount a plow on torsion bars unless it was for personal use. With the torsion bars, I would get the lightest blade I could find."

Ultimately though, the "right" specifications for a vehicle depend on what you use it for. "If I intend to plow mainly large commercial lots," Smith said, "I would want a full-size vehicle, maybe a 1-ton dump truck. Or, maybe you will be plowing with your one daily driver pickup, so the best choice would be a ¾-ton truck. That's an example of how everyone has different needs."

ODDS AND ENDS. Once you've tackled the big questions and selected a truck that fits your unique needs, there are a host of other miscellaneous items to consider. For instance, McTavish suggests spending an extra $300 to $500 to have electrically operated windows and heated mirrors. These options provide greater visibility for drivers, and the electronic controls allow a driver to easily adjust settings to his preference, which is especially important in snow removal since several different drivers may use the same truck throughout a storm.

Heated windshield wipers are also available as an optional upgrade to improve visibility. "I think having those features adds a lot more safety – a lot more than you think," McTavish remarked. "With all our vehicles we try to make it as safe as possible for the guys to drive them."

McTavish also advises contractors to pay attention to the ergonomics of the cab, since you're likely to be spending a significant amount of time in the vehicle. "Someone that sits in a vehicle quite a bit, you realize the small things you might not notice when you buy a vehicle," he pointed out. "For example, leg room is important. When your foot is on the gas, instead of your knee being shifted off to the side, you need to have enough room to keep your leg straight or it may lead to knee problems."

CONCLUSION. There are a wide variety of vehicle options available to snow removal contractors. Each option has its benefits, disadvantages and associated costs. The best strategy is to take a good look at what you need, ask yourself what you want the vehicle to accomplish and find the truck that can handle your requirements.❄

Reprinted with permission from Snow Business *magazine.*

equate for residential plowing situations, they often don't do well in commercial plowing operations.

Straight blades are less expensive than V-blades, but they have limitations on what they can do efficiently. The straight blades must be angled in order to be productive. Pushing "straight on" with a straight-blade plow is the least efficient method of moving snow. When you add a set of snow wings to the conventional straight blade, you will increase efficiency by 35 percent, but you are adding weight to the front end of the truck.

Deciding on a V-blade will cost you as much as $1,000 more than the conventional straight blade, but it is usually twice as efficient as the straight blade, and well worth the added expense. V-blade plows take more talent to operate, and it can take as long as a season to get used to the controls so that you can operate the unit without thinking about it. But, again, the increased efficiency makes the slight inconvenience worth it.

Let's Make A Deal. The most important thing to consider when purchasing a plow for your truck is the dealer support you can expect. Some dealers will promise the world and then not have time to spend with you when there is a problem. Buying a cheap plow can become quite expensive if you cannot get replacement parts or if the dealer is only open a few hours during a snow storm. Ask the dealer for references. Ask other snow professionals about their experience with various plow dealers. Having a great dealer to work with can offset the expense of the initial purchase.

Don't Trip Up On Trip Blades. A "Trip" feature on a snowplow blade allows the plow to give when passing over raised objects, such as manhole covers. The feature is typically available as full-trip or bottom-trip blades. Ask a plowing contractor who has been using bottom-trip blades his entire career, and you'll find that the bottom-trip plow is the best. Ask the same question of a plowing contractor who has used nothing but full-trip blades, and you'll usually get the same type answer. One is not better than the other.

Often the market area drives the subject. One manufacturer is very strong in the northeast, for example, and its products are the products of choice in that market area. Head to the upper Midwest and another manufacturer has a very strong presence there. Thus, its products are the ones everyone sees and buys. Check out the dealers and plows in your area and decide which is best for you.

Putting A Push On Snow. One new style of snowplow, the box

NOTABLE SNOWSTORM: February 1977: Ontario, Canada and western New York were slammed by a storm that killed 28 people and shut down the city of Buffalo for more than a week. Highways were clogged with thousands of stranded vehicles, and people became trapped at schools, stores and offices, where they were forced to spend the night because they could not make it home through the blizzard.[3]

plow (sometimes called the push plow or a snow pusher), originally came to be used in the Rochester, N.Y., area more than a decade ago. As the units became more recognized and used more, engineers designed better and better units. Once patented, snow pushers became more commonplace in the snowbelts of New York and Pennsylvania.

Many contractors have been drawn to these box plows because they can remove more snow per pass than traditional plows. This increases efficiency, decreases plowing time and allows the contractor to serve more accounts – and that means more revenue.

Several brands of box plows exist in the market today in a variety of sizes and quality levels. The best box plows are constructed of heavy-duty steel with a properly engineered support structure. Such units can move hundreds of cubic yards of snow in a single pass. And in deeper snows their efficiency increases even more dramatically.

Box plows should be used on loaders only. Some might argue box plows can be utilized on trucks; however, extreme caution should be used in such instances. The box plow is, by nature, constructed very solidly. It must be mounted directly to the frame of a truck (if truck mounted). Because of the solid construction, any energy shock from a front-end collision (even with a manhole cover) is transferred directly to the frame of the truck. In some of these instances there will be a failure of some structure, most likely the frame of the truck. This is an expensive repair.

As noted above, box plows are for use on loaders. They come in all sizes, from 6-foot models for use on skid steers to 30-foot-wide models that are used primarily in airport applications. Box plows are easy to mount on a loader and will increase efficiency greatly. In per-push or seasonal pricing situations, they can increase margins dramatically, pay for themselves in the first year of use and are becoming the equipment of choice on large properties.

Poly Edges. Another recent advancement has been the use of polyurethane cutting edges on plows and pushers. Numerous plowing contractors are replacing steel cutting edges with polyurethane when the steel wears out.

The reason for the revolution is that the polyurethane cutting edges are saving contractors money. While the initial expense of the polyurethane edge is several times that of steel edges, polyurethane will last as much as five times longer than steel. When you see a plow truck going down the road with sparks flying, it is the steel cutting edge coming apart or wearing. Polyurethane edges don't come apart

Today many companies successfully use "heavy equipment" for snow removal. Pictured (left) is a loader with a box plow attached and (right) a skid steer with a straight-blade plow.

and are hard and thick enough to retain some cutting ability. Additionally, due to the sliding action, a polyurethane edge will not disturb gravel parking surfaces nearly as much as steel. Having to redistribute gravel after the snow melts can be a bone of contention between property owners and contractors. Polyurethane virtually eliminates these problems.

Probably the most important reason for using polyurethane is that it is very forgiving. Imperfections in the parking lot surface will not be disturbed by polyurethane, while steel may peel up the pavement when an imperfection arises. Manhole covers, cracks in the pavement, grates that stick up slightly and pieces of cracked concrete are not disturbed by polyurethane. And, the polyurethane will "remember" its original configuration and return to its straight edge after bending to go over such imperfections in the plowing surface. Eventually, polyurethane will be the cutting edge of choice and steel will be the option in order to reduce the cost of the initial purchase.

USING HEAVY EQUIPMENT

Long gone are the days where snowplow operators relied solely on pickup trucks to remove snow. Today, many successful companies use a wide variety of so-called "heavy equipment" in snow removal operations. The most commonly used units are skid-steers, loaders and dump trucks. Often this equipment is already available to a snow removal contractor from his summertime business, particularly in the case of landscape and construction industry professionals. Many contractors also successfully form subcontracting agreements with companies who can supply this heavy equipment and the appropriately trained opera-

SMART SNOWPLOW SELECTION

Snowplow manufacturers have been working hard to develop new plow technologies that save time, make the plow easier to handle and maneuver and, in general, make plowing faster and easier. Conditions are rarely perfect when plowing; manufacturers of snowplow equipment must consider the variables and incorporate solutions for a number of situations. Contractors should be able to make a smart plow purchase once they are educated in the differences between plows, the important features to look for when buying a plow and the best technologies available from snowplow manufacturers.

STRAIGHT VS. V-PLOW. Snowplowing technology made a huge advancement and revolutionized the snowplow industry with the introduction of the V-Plow more than a decade ago. Although the straight plow is still available and widely used, the V-plow is quickly taking over as the preferred choice in snow removal equipment among professional snowplowers.

The V-plow differs from the standard straight-blade plow because it is made up of two halves, or wings, that can be independently adjusted to multiple positions based on snow conditions, plowing location and the individual operator's needs.

When both wings of the V-plow are in the forward position, the plow forms a scoop that carries more snow and controls the snow better than a straight-blade plow. A straight-blade cannot achieve the scoop position, and because of a straight moldboard, the snow has a tendency to spill over the ends of the plow.

The wings of the V-plow can also be retracted (toward the truck) to form more of a V shape which will push through deeper snow more easily and with less stress on the vehicle. A straight-blade plow requires more power from the vehicle to push through deep snowdrifts and often results in the vehicle sliding sideways on the surface, rather than pushing through the snow.

It is important to note that a V-plow can also function like a straight plow. Because the wings are multipositional, they can be positioned straight across the front of the vehicle when the conditions are suitable for this type of plow.

Straight-blade plows are often used in an angled position to windrow large, obstacle-free areas. This type of plow is effective in large parking lots or other open areas because the plow pushes the snow to the side in rows, one blade-length at a time. The snow that is pushed to the side piles up on top of unplowed snow and all of this snow is plowed on the next pass.

Straight-blade plows also tend to cost less and weigh less than the V-plows.

When shopping for a plow, whether it be a V-plow or a straight-blade plow, there are a few important features to look for, including the size, strength and durability of the blade, the blade trip, the controls, the undercarriage system, the mounting system and the hydraulic system.

SIZE MATTERS. Most often the bigger the plow, the bigger the truck needed to carry it. Snowplow manufacturers take this into consideration and most offer a wide variety of sizes. Plows can range in size from 6½ to 11 feet wide. Compact to mid-size trucks will use a plow of 7½ feet or less. Half-ton trucks will have plows ranging from 7 to 7½ feet, and three-quarter-ton trucks and larger will use plows in the 7½- to 11-foot range. Having the wrong size plow on a truck can mean excessive wear and tear on the vehicle and reduced plow performance. A plow that is too heavy or too large for a truck may overload the front axle and present safety issues like reduced braking and steering control.

STRENGTH VS. DURABILITY. Snowplow blades can be made of steel, which is the most common, or plastic. Depending on the design, features on steel-blade plows come in various strengths. For example, high-strength steel may be used for the push frame and push beam hook, while a mild- or common-strength steel will be used for the actual blade or center section. Plows designed with high-strength steel are stronger, more durable and react better on unpredictable surfaces.

The steel blades can be finished by paint or powder coat, or a baked-on, two-part epoxy paint. Care has to be taken with painted and powder-coated finishes so they don't chip, as this can lead to rusting and/or rust bubbles, which will shorten the life of the plow. A powder-coat finish consisting of phosphate pretreat and TGIC polyester provides a hard, high-gloss surface that is smooth and slick. A UV-resistant finish is also important to ensure the color won't fade or degrade when subjected to the ultraviolet rays of the sun. There are two types of plastic snowplow blades – clear and color-impregnated.

Plastic plows have their advantages and disadvantages. Plastic has a lower coefficient of friction and lower moisture absorption so the snow tends to roll off more easily. This can result in less fuel consumption when plowing. Plastic plows won't rust or dent and can be replaced relatively easily if necessary. Plastic is also lighter so some models do offer weight savings. However, this is not always the case, as the plastic blades are not as strong as their steel counterparts and may require additional bracing.

THE TRIP. The blade trip is another important feature because snowplowing often occurs under conditions of poor visibility and in areas with unpredictable surfaces. The blade trip is designed to assist in these tough conditions with the use of a spring-loaded system that causes the moldboard at the bottom of the blade to pivot forward and bounce, or "trip," over unmovable obstacles.

There are two basic styles of tripping blades – a full moldboard trip and a partial moldboard trip. A partial trip means just the cutting edge up to 6 inches will trip and bounce over the obstacle. If the obstacle is more than 6 inches tall, the moldboard might not go over. A full moldboard trip means the entire blade will pivot when an obstacle is struck. This allows the plow to easily maneuver even in the most uncompromising conditions.

CONTROLS. Operator fatigue is common since contractors often work all hours of the day and night through changing and unpredictable weather conditions to ensure roads are safe and driveways, parking lots and sidewalks are accessible. It is important they have complete control of the plow and can make changes and adjustments to it at a moment's notice. There are four types of controls: switchbox, joystick, touchpad and handheld.

The switchbox is relatively easy to use, economical and reliable but rather unattractive. Joysticks have the ability to operate more than one function at a time and are easy to use, but they need to be securely mounted inside the cab. Touchpads and handheld units do not require mounting and vary in functionality and ease of use.

UNDERCARRIAGE STRUCTURE. Because the V-plow is new to plow manufacturers, it is important to take a look at the undercarriage used to mount the plow on the vehicle.

Most manufacturers have developed adequate undercarriage structures for the straight-blade plow. However, with the advent of the V-plow, most manufacturers have not yet developed mounting systems specific to this type of plow, and continue to use the systems designed for the straight-blade plow. This method is not as effective. The plowing forces are much greater on a V-plow, therefore, the V-plow requires a stronger and more advanced mounting system.

MOUNTING SYSTEMS. Most of the newer mounting systems allow the operator to drive right up to the plow vs. the older, more difficult way of manually lifting the plow with another piece of equipment. The drive-up removable mount system means the vehicle can be restored to normal when not plowing.

Many plow manufacturers have mounting systems that still rely on the use of locks and pins, and some require two people to complete the mounting process. However, newer systems are becoming available that do not involve pins, locks or other removable parts that can get lost in the snow.

HYDRAULIC SYSTEMS. There are two types of hydraulic systems, and it is important to consider which system is appropriate for the plow and the vehicle. The belt-driven hydraulic system is driven off of an engine-mounted belt for power up/down. This type of system can cause an additional draw on the engine and can affect vehicle emissions. This system is slowly being phased out, as mounting them under the hood is increasingly difficult because of limited space.

The more common hydraulic system is run by the vehicle's electrical system via an electric cable reel system, which is usually a power up/gravity down system. Advances in DC motors have reduced the electrical load and made this the standard system in plowing. It is important that the mounting system/hitch has a consistent down force to keep the plow on the ground.

When considering electrical systems, it is important to look for heavy gauge electrical cables that go to the electric motor to minimize electrical power losses. For commercial applications, a second battery may be added to handle the demands of plowing for long periods of time. It is also important to look for hydraulic systems that are shielded from outside elements such as moisture, dirt and corrosion. These shields will protect the system form damage and wear. ❄

Reprinted with permission from Lawn & Landscape *magazine.*

tors, such as residential/commercial builders.

Using or subcontracting the use of this heavy equipment can greatly boost a contractor's snow removing efficiency. Such equipment can be mounted with large plows or snow throwers that can move substantial amounts of snow in a relatively short amount of time. However, this equipment does require extra training and maintenance.

The proper way to use such equipment is a book in itself. Suffice it to say using heavy equipment to remove snow on large, commercial customers can significantly boost profit. However, make sure you are well-versed on the proper, safe-operation methods of such equipment, as well as the ongoing required maintenance.

There are instances when smaller, more maneuverable equipment like power brooms (left) and snow throwers (right) are called for.

SWEEPERS/SNOW THROWERS

There are instances, clearing sidewalks for example, when smaller, more maneuverable is called for. In those instances contractors often use walk-behind snow throwers, tractor-mounted sweepers or three- or four-wheel all terrain vehicles (so-called ATVs or Quad Runners) mounted with plows.

Such items allow contractors to quickly and efficiently clear snow from otherwise inaccessible surfaces. Such equipment is much faster and often much more profitable than using crews to physically remove the snow with shovels.

DEICING SPREADERS

Deicing will be discussed in greater detail in Chapter 14. However, some basic discussion of deicing equipment is warranted. Deicing equipment can range from the simplest hand-powered spreader to a sophisticated, electronically controlled liquid tank and application system mounted on a commercial truck chassis.

The most commonly used deicing equipment for private snow removal professionals is the back-end-mounted hopper/spreader or the slip-in V-box spreader. A wide variety of sizes and holding capacities exist within these product categories, which can be selected by the contractor based on budget and spreading requirements. Price can also vary depending on the mechanism used to control the rate at which deicing material is spread. The most convenient units have electronic controls mounted inside the cab of the vehicle. The hopper spreader is mounted on the bumper of a pickup truck and can typically hold around 9 cubic feet of material. The V-box spreader is mounted inside the bed of the truck and therefore can hold more material. Load capacity is typically determined by the size and weight of the vehicle in which the V-box will be mounted and varies from 1 to 7 tons.

❄ SNOW THROWER SAFETY TIPS

Each winter, emergency rooms in snowy parts of the nation routinely see injuries to hands and fingers from snowblowers. Make sure you avoid such injuries this winter by following these tips:

- If the chute becomes clogged, turn off the snow thrower, wait for all moving parts to stop, and then, with a stick, clear the chute. Never put your hands inside the chute for any reason.
- Be careful never to throw snow toward people or cars, and never allow anyone in front of your snow thrower.
- If you have to repair your machine, always turn off the snow thrower, wait for all moving parts to stop and disconnect the spark plug wire.
- Keep hands and feet away from all moving parts.
- Dress properly for the job. Be sure to wear adequate winter garments and footwear that will improve footing on slippery surfaces. Wear safety glasses and avoid any loose-fitting clothing that could get caught in moving parts.
- Handle gas carefully. Avoid spillage by using non-spill containers with spouts. Fill up before you start, while the engine is cold. Store gas in A clean, dry, ventilated area, and never near a pilot light or heat source. Never smoke around gasoline.
- Do not clear snow across the face of slopes. Use extreme caution when changing directions on slopes. Do not attempt to clear steep slopes.
- Never operate the snow thrower without good visibility or light. Always be sure of your footing and keep a firm hold on the handles. Walk, never run.[4]

TRUCK & PLOW MAINTENANCE

Equipment maintenance costs can easily get out of hand if you ignore preventive maintenance issues. Being proactive instead of reactive can save you thousands of dollars in emergency repairs. It is recommended that a formalized preventive maintenance plan be instituted. This can be as simple as posting guidelines on the calendar in your computer so that reminders pop up regularly. For example, it is easy to post that a regularly scheduled lube/oil/filter change is due.

Oil Changes. In the winter, it is highly recommended that you do a lube/oil/filter service every month. This might seem excessive, but keep in mind you are working the truck very hard when plowing snow. Treat the truck right and it will last much longer. The only exception would be operators with light snow accumulation. Obviously, as they might only use their snow removal truck once during the winter, they wouldn't need to change oil monthly.

Typical deicing spreaders include a rear-mounted spreader (top) and a V-box spreader (bottom).

Transmission Fluid. A snow professional should change the transmission fluid more frequently than stated in the manual (approximately every 12,000 miles), especially if he has come through a particularly hard snow event or a series of snow events. Transmissions get hot during plowing operations. Fluid burns when it gets hot and loses its viscosity. Change the transmission fluid often and the transmission will last much longer.

Hydraulic Fluid. Any leaks in the hydraulic system of the plow unit can be cause for concern. Even as fluid leaks out, water can seep in where the leak occurs. Leaky piston seals, couplers, connectors, etc. can all allow water into the system. The water will settle to the bottom of the reservoir and can be easily drained, however, the water can cause rust to form on internal parts and surfaces. This is not good for the system and should be addressed as soon as noticed. And, it is advisable to change the hydraulic fluid at the end of the winter season so you can start with fresh fluid in the fall.

Grease. Grease is very important. Universal joints should be checked every two snow events and greased as needed. The hinge joint on the V-blades should be greased after every plowing run. Same for the hinges on the fold-out snow pushers.

Any other grease fittings on plowing equipment should be checked regularly. For loaders and skid steers, fittings should be greased daily. It cannot be stressed enough just how important this task is to the continuing good running order of the equipment.

Salt Spreading Maintenance. Preventive maintenance on salt spreading equipment is also very important. The corrosive nature of salt demands that care be taken with moving parts in salt spreaders. The spreaders should be thoroughly cleaned after each use.

After cleaning, the application of a fine film of an anti-corrosive mixture will go a very long way toward keeping moving parts in working order. Some contractors use a mixture of diesel fuel, water and dishwashing soap (60/30/10) sprayed from a one gallon pump sprayer on all surfaces that come in contact with salt as a method of protecting these parts from corrosion. Only a few ounces of liquid material is required to adequately cover all surfaces. This also allows the salt to flow smoothly out of the hopper.

Your equipment is expensive both at the time of purchase and when you must make repairs. Most contractors know and understand that preventive maintenance costs much, much less than reactive maintenance, such as major repairs.

REPLACING EQUIPMENT

Contractors constantly struggle with the question of just how long to keep equipment and trucks in service before retiring or replacing. There is no hard and fast answer.

Cash flow often enters the decision-making process. There is no question that an owner-operator of a plow truck will take considerably greater care with its upkeep than the typical paid employee. A well-maintained truck (even with lots of plowing, operated by an employee) should last at least five years. Ten years is easily obtainable with a proper preventive maintenance program and with controls in place to properly train operators.

One suggestion would be to keep track of all operational and maintenance costs for the vehicle in question. If the combined costs in a given year exceed the payments for a new vehicle (and its operational costs) then its time to consider replacement.

Plows are a different story. It is known that the operational lifetime of a plow is around six or seven years, maximum.

This cost is usually much easier to consider when projecting cash flow. Salt spreaders will last eight to 10 years (excluding the engine on those units using engines). Hydraulically operated spreaders will last considerably longer if well maintained. ❄

SOURCES

[1]U.S. National Weather Service

[2]National Snow & Ice Data Center, University of Colorado, Boulder, Colo., http://nsidc.org.

[3]National Oceanic and Atmospheric Administration, National Snow & Ice Data Center, University of Colorado, Boulder, Colo., http://nsidc.org. AccuWeather.com, www.accuweather.com, National Climatic Data Center.

[4]Outdoor Power Equipment Institute, Inc.

CHAPTER TWELVE: Plowing Effectively

IN THIS CHAPTER

- ❄ Using the right equipment for the job
- ❄ Efficient snowplowing techniques
- ❄ Plowing safety tips
- ❄ Avoiding costly mistakes
- ❄ Training others to plow effectively

Those who have been plowing snow for several years have developed systems for plowing that are quite adequate, and it would be presumptuous of the author to think he can improve upon age-old methods for moving snow. However, some suggestions might assist novice plowers on how to better manage the snow on the sites they service.

Those who plow snow on a per-hour/per-truck basis have little incentive to move snow on customer sites efficiently and effectively. Experience shows that those who plow snow on a per-push basis inherently seek out the most efficient way to clear the snow off a particular site in order to move from one site to another as quickly as possible, so as to increase production times and profits. For purposes of this discussion, we shall assume the plowing contractor wishes to increase production times so as to increase the aforementioned revenues.

MATCHING EQUIPMENT TO THE JOB

We have previously discussed the use of proper and efficient equipment in the moving of snow from pavement surfaces. However, the mindset of how to best use that equipment becomes important when attempting to increase efficiency. Think outside the box when detailing how the crew should "attack" a site during a snow and ice storm.

Part of the problem is that, once a mindset is established, it becomes quite difficult to go to the next level. Be vigilant in always considering more efficient ways to clear snow from an ac-

Matching the right equipment to the job, from a simple shovel to heavy equipment, is crucial to ensure efficiency.

Always consider the most efficient means to remove snow. It may take hours to clear this lot with a walk-behind machine, whereas a box plow might be able to clear the lot in minutes.

count. For example, most contractors who finally use a snow pusher are amazed at the increased productivity. However, these same contractors will continue to plow smaller lots with pickup-mounted plows instead of trying to use the pushers to increase productivity across the board. Try to suggest that a 20-foot-wide, fold-up snow pusher be used in a typical fast-food restaurant, and the closed-minded person will think this is a crazy idea. In fact, this is being done now in several markets. Imagine plowing the fast-food restaurant in 10 minutes using a 20-foot pusher and a plow truck. Now imagine the profit that results if you are charging per push.

The plow truck pulls snow out of the tight corners into the main lot, and then leaves the site. The loader and 20-foot pusher come in and make two passes around the building, totally clearing the lot.

The use of a 10-foot pusher on a four-wheel-drive (4WD) backhoe will also do the same fast-food restaurant all by itself – assuming an open-minded and capable operator is at the helm. Pushers on 4WD backhoe loaders will backdrag, get close to curbs, tuck corners and pile snow, just like plow trucks, only much faster.

CLEARING SNOW

Let's assume you are not utilizing snow pushers yet, and you are using plows on trucks. In parking lots where the lot completely surrounds the building, the old-fashioned method would dictate you plow each side of the building in turn. When diagramed, one can ascertain that this is an effective method of plowing.

One moves the snow from the lot on one side of the building, then clears the second side by backing up onto the already cleared corner of the lot. This is repeated around the building until the entire lot is cleared. Unfortunately, this method results in considerable backing up of the plow truck. Going forward as much as possible, with the plow down and pushing snow off the pavement, is preferable. However, the old school thinking won't allow for any other ideas to come into play.

How else can you clear the lot if you don't push the snow to the outside edges one side at a time? Envision, if you can, the lot is clear of vehicles. Also, envision the interior nooks and crannies are "pulled out" into the lot first. One pulls the snow out past any islands or other close-in obstructions prior to doing any forward plowing.

Once the backdragging is completed all around the building, you then begin running around the building in circles (counterclockwise in this example) angling the blade to the passenger side of the truck.

With each concentric circle around the building, you go in wider and wider circles, pushing the snow closer and closer to the perimeter of the parking lot. Eventually you may have to go around light poles in the lot, but by cutting away from the base of the pole, you'll eventually get out past that pole and continue your concentric circles, further advancing the snow toward the perimeter of the lot.

Once the vast majority of the snow is at the exterior curb lines, you then push in the entrances to the lot (from the street), and tuck the corners to neatly finish off the plowing.

Experienced plowers will adamantly tell you that you'll cut the plowing time by almost 50 percent in this fashion. Exact time reductions will depend on the number of obstructions in the lot as well as ongoing snowfall restrictions. However, dramatically reduced plowing times will result if the above example is followed closely.

The above example does not work in all circumstances and, often, one will find themselves doing the plowing in the most efficient manner anyway. But, if you can learn to think "outside the box" when approaching the commercial lot, you'll be amazed at how you can increase production.

When plowing, make sure you know how and where the customer wants snow piled.

WHEN ARRIVING ON SITE

Some examples of things you must be cognizant of when entering a lot to begin plowing operations are as follows, in no particular order:

- Ascertain where catch basins

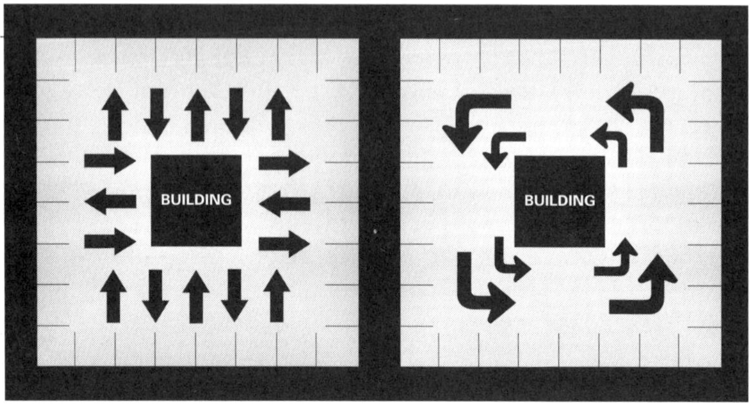

A circular rotation can be more efficient than the traditional forward and reverse method of snow removal.

are located so you don't inadvertently create a dam that might cause flooding.

- Be aware of curb locations so you don't damage your plow (or the curb) during plowing operations.
- Know where fire hydrants are located and never plow snow onto a hydrant. Firefighters may know where the hydrant is located anyway, but making them dig out the hydrant in an emergency slows down the response time in the event of a fire at the site.
- Don't plow snow onto sidewalks intentionally. Most walks are to be kept clear for pedestrian traffic.
- Be aware if the customer allows plowing snow onto shrub beds or planted islands.
- Watch for mperfections in the parking lot that can cause head banging, so-called"wake up calls," if struck while plowing.Not only does this create severe headaches, but can cause injuries that require stitches. A visit to the site prior to the onset of winter will allow you to know where these are located.
- With automatic transmission vehicles, always plow in "low" on the column, even if the run is short. This does not mean plowing in 4WD Low. Do not plow in "drive" as the transmission will attempt to shift gears as the resistance to the truck increases or decreases during plowing operations. You can quickly burn out the transmission or transfer case on your truck by not plowing in low.
- Always disconnect overdrive switches when plowing. Never plow with overdrive engaged. You can burn out the transmission in minutes this way.

SNOW TERM: SNOW SQUALL

A brief, but intense fall of snow that greatly reduces visibility, which is often accompanied by strong winds.[1]

One Very Important Cautionary Note: Always plow with the seat belt securely fastened. It is inevitable you will hit an obstruction of some sort. It may be a piece of pavement sticking up or a manhole cover that is up a half inch. The truck may come to a dead stop in such instances, but the physics of inertia will allow your body to continue its directional path. It usually takes one time of kissing the steering wheel to reinforce this edict.

TIPS FOR A QUALITY JOB

Some important tips for providing a quality job for the customer include:

- Clean up after yourself in the lot. Don't leave "feathers" or "trailers." A clean lot is what most customers expect even if you have not specified this in the contract. Lines of snow from feathers make the lot look messy and give the impression the plower doesn't care about quality.
- In lots that don't completely surround the building, try to plow in straight lines. It makes for a much neater appearance and implies professionalism.
- Cut corners tight and square them off. Just veering off at the end of the run without squaring off the corners gives the appearance of not caring about the finished product.
- At the entrances to the lot, refrain from piling up snow at the corners of the apron. This can restrict vision when cars attempt to exit the lot. You could be held liable for creating a dangerous situation if an accident occurs.
- Don't intentionally "plow in" cars that may be parked in the lot. Even if the car shouldn't be there, you may be considered liable for damages inflicted.
- Keep parking lot aprons free of excess snow so vehicles may easily enter and exit from the lot safely.

PLOWING SAFETY

Snowplow operators are asked to work in very hazardous conditions, often late at night and in difficult circumstances. As a result, injuries, vehicle accidents and damages to the customer's property can occur if safety is not a top priority. Also, being unprepared for equipment breakdowns can create dangerous situations in which the plow operator is exposed to the elements or vehicle traffic while trying to fix the broken machinery. Some important safety/preparedness tips when plowing or getting ready to plow include:

- Check all fluids regularly, including the plow fluid, engine oil, transmission fluid and washer fluid.
- Check tire pressures regularly. An overly hard or overly soft tire can hamper efficiencies.
- Carry spare pins, hydraulic fluid and a spare hydraulic

NOTABLE SNOWSTORM: February 1978: "The Great New England Blizzard" dumped heavy snow on much of New England and the eastern portion of the Midwest.[2]

line in case one blows during plowing operations.

- Carry tools necessary to change a hydraulic line, or to add fluids, or to even effect minor repairs out in the field.
- Carry spare clothes just in case you have to get down under the vehicle while it's snowing and you get wet. Getting sick is not an efficient way to continue generating income while it's snowing.
- Carry a roll of toilet paper. Trust me on this one, you'll be very happy you have it.
- All plowers are convinced that they will never get stuck. However, in the unlikely event it happens to you (or some other poor sap), carry a tow chain.
- Have gloves with you. You'll be very glad you do if you ever get stuck during a storm.
- A few bungee cords can come in handy often. Place a couple under the seat "just in case."
- Fluids, such as water or a sports drink, can be important during long stretches in the truck. Obviously, beer or any alcohol is never appropriate while operating a vehicle or piece of equipment.

SNOW FACT: The old saying that 10 inches of fresh snow contains 1 inch of water is only occasionally true. In reality, 10 inches of snow can contain as little as 0.10 inches of water to as much as four inches.

AVOIDING MISTAKES

Here are some common mistakes plowers make when starting out. Unfortunately, some of the most experienced plowers can make some of these mistakes too.

- Taking too big a bite when plowing. This allows for feathers to develop.
- Trying to stack the snow too high with the plow. This can damage the plow frame. It can also cause the front axle of the truck to get hung up in the snow pile, requiring the use of that chain you didn't think you would ever need.
- Not asking for help when/if you do get stuck. It is amazing just how much camaraderie there is amongst snowplowers when they are out there fighting the storm. Assistance when you are stuck is just a short distance away.
- Not taking a moment to ascertain the most efficient plowing pattern for the lot you are in. Just digging in and getting it done can lead to double plowing and inefficiencies in your pattern.
- Being in too much of a hurry. Remember, you're never too busy to do the job right. Customers get mad if you do a hurry-up job. We're professionals and should treat every job and every site as if it is our most important project.
- The job is never done until the paperwork is completed. Keep track of what you do and write it down. After a long stretch in the plow truck it is often difficult to remember what was done in the

heat of battling the snow and ice storm.

- Not documenting conditions while you are plowing. Inevitably there will be insurance issues that arise down the road (and it could be years down the road). Trying to reconstruct a snow event 18 months later is difficult.
- Don't panic when it is snowing hard. If necessary, take a deep breath, review the situation, make a plan and then follow it.

TRAINING

Experience is the best educator when it comes to plowing snow. Being efficient and methodical can make life much easier during a snowstorm. Unfortunately, as your company grows, you will be less and less involved in the actual plowing of accounts, which means you will have to learn to rely on your employees. This means you will need to conduct adequate training, so that employees serve your customers to the level of quality you expect.

Equipment operators require training in order to properly use snow removal equipment. Instruction and training are key ingredients to success in the plowing business. For example, instructing the loader operators that their buckets need to be in the float position and not to attempt to apply down pressure to the rear of the pusher, is key to their ultimate success in operating these units. The tendency for any loader operator is to apply more down pressure in order to cut into the snow, however, these units are designed to slide over the surface to adequately remove the snow in front of the unit.

Training resources, such as these videos from the Snow & Ice Management Association, are available to snow professionals.

Plow truck operators also require training in order to adequately perform. Never assume that the new rookie plow driver instinctively knows how to plow snow. Start by showing a proper training film that details the proper pre-storm preparation, proper methods of plowing smaller properties, and what to watch out for when plowing on larger sites. You may also want to establish a formalized training program. ❄

SOURCES

[1]National Snow & Ice Data Center, University of Colorado.

[2]National Oceanic and Atmospheric Administration, National Snow & Ice Data Center, University of Colorado, Boulder, Colo., http://nsidc.org. AccuWeather.com, www.accuweather.com, National Climatic Data Center.

CHAPTER THIRTEEN: Sidewalk Snow Removal

IN THIS CHAPTER

- ❄ Sidewalk snow removal as a profit opportunity
- ❄ Efficiently clearing sidewalks
- ❄ Sidewalk snow removal crews
- ❄ Small crew vs. large crew

Another area you will be forced to consider is sidewalk snow removal. If you don't already offer this service, a customer will likely eventually ask you to provide it. To be profitable, this area requires attention to detail, as well as attention to productivity goals.

Further, sidewalk snow removal probably demands the most management time and effort. However, this area can generate higher margins than the snowplowing portion of the business. It takes good organization and goal setting to achieve maximum performance from your crew(s) and to achieve maximum profits.

Managing performance in snow and ice management organizations (whether in on-site situations or multi-location accounts) is a complex issue. With regard to sidewalk snow removal, it involves all of the components of productivity. In most organizations, people are hard-working, or at least they believe they are. Announcements of productivity goals or productivity increases will most likely be interpreted by your people as a

demand for the frontline worker (specifically the sidewalk snow removal workers) to work harder. This is not true. When introducing sidewalk snow removal into your offering of services to be provided to customers, we want the crews to work effectively and be accountable for the results of their

To be profitable, sidewalk snow removal requires attention to productivity levels.

Sidewalk snow work is a labor-intensive activity.

efforts. This is true not only from the standpoint of the contractor, but also from the viewpoint of the customer.

People are, by nature, territorial. We tend to accept responsibility and be more accountable when our territory is defined. Often it is difficult for the snow removal laborer to accurately perceive the scope of "his territory," because in snow and ice management the areas are so large and diverse. It is also difficult for management to track performance.

THE SIDEWALK SNOW REMOVAL LABORER

Sidewalk snow work is a labor-intensive activity. Most payroll dollars go to production people as a percentage of the whole labor dollar outlay. The labor intensity of sidewalk snow removal has improved in the past and the percentages may change in the future, but people power will be the heart of the business for years to come.

We have not been able to substantially reduce the man-hours per lineal foot of walk, even in the light of improved ice control materials and more sophisticated equipment.

Further, the U.S. Department of Labor has confirmed our fear that the pool of available workers is shrinking and will get even smaller in the future. The industry will continue to experience more companies chasing fewer available workers. Unfortunately, since our customers are demanding this service be provided, we will have to find ways to accommodate them.

We need to evaluate the importance of the production worker. It is difficult, at first, to accept the reality that the production worker in snow removal operations (often the lowest paid and generally an on-call, part-time employee) should be the focal point of our sidewalk management systems.

A production unit for sidewalk snow removal is a crew that includes labor, equipment, material and transportation. The labor for one production unit usually consists of one crew leader (or working foreman) and one or more crew members.

The Crew Leader. The crew leader has emerged from the 1990s as the specialist of the new century, with an expanded role in managing the snow removal and ice control on any specific site. The difficulty and expense of communication links and direct supervision of mobile service crews, coupled with the need to have an experienced employee on the property at all times (as is the case with most on-site crews) has reshaped the value and job description of the traditional working foreman or crew leader.

Organizations that recognize this expanded role for the crew leader will streamline their organization by eliminating the middle managers and production supervisors. They will redistribute these responsibilities all the while upgrading the role of the modern day crew leader.

The Production Leader. Most snow-related production workers who are attracted to our industry like to work outdoors, or they are doing snow removal as part of a year-round activity that includes landscape or property management. Normally they do not object to physically demanding work.

However, we must provide them with the proper tools to do their job, as snow-related tasks are (by nature) much more physically demanding than landscape maintenance, pavement maintenance or equipment operator duties. They may need a pay differential for winter-type work, which often means working in conditions that are intolerable to others in the work force.

What specific traits should we look for when hiring a production person to work under an experienced crew leader? After the basic requirements have been met, look for the person who needs to be active at all times and who appears to be bursting with energy. The production worker we look for usually has little regard for detail and is difficult to train in the classroom. He learns by doing and is a good candidate for on-the-job training, provided the objective is getting him to take action, not improving his knowledge.

The sad reality is most true production people, on sidewalk snow removal crews, leave our

SNOW FACT: Though relatively rare, thunder and lighting can occur during snow storms, especially lake-effect snows. Lake-effect snows are caused by temperature differences between relatively warm lake water and a cold upper atmosphere. If the temperature contrast is great enough, the rising air will have enough buoyancy to form "thundersnow," which are thunderstorms that have snow falling out of them rather than rain.[1]

NOTABLE SNOWSTORM: March 1993: The "Blizzard of the Century" ravaged the southern mid-Atlantic states from Alabama to Massachusetts, accompanied in other states by severe weather disturbances such as tornadoes, thunderstorms and floods. Snow fell at rates between an inch and 2 inches an hour in some areas, and many locations experienced record-breaking snowfalls and record snow depths. The storm caused approximately 270 deaths and $3 to $6 billion damage.[2]

industry because they do not see themselves as managing a production operation. This is true even in other service industry operations as well as snow removal operations.

The Production Team. A production organization (large or small) should be staffed and organized so the entire operation is a support system for the production workers and production units. Sidewalk snow removal should be viewed as a production-oriented task. The functioning production organization should be structured so management will not interfere with production.

Production workers perform best when they are managed as team members, or athletes, rather than laborers. They need specific goals set for each production period (or snow event).

The company standard (either the contractor's standard or the customer's standard) for performance must be demonstrated by the crew leader. The procedures leading to the standards must be taught while a snow event is taking place, thus putting even more pressure on the crew leader and production workers.

Unfortunately, snow events in some areas of the country are few and far between, thus making retention of the production principles even harder.

Most people want more than a paycheck for a day's work. Part of management's responsibility in a production organization is to help each player on the team to build self-esteem.

SMALL CREW THEORY

The most efficient crew size for performing sidewalk snow removal activities has been discussed, argued and subjected to trial-and-error testing. Since sidewalk snow removal emerged as a separate (or specialty) business, the issue became more important. By adding mobile crews, we discovered the importance of correct crew sizing. In today's competitive labor environment, the need for higher productivity and increased quality suggests a new look at sizing sidewalk snow removal crews.

Let's look back at our (non-snow related) experience with small crews. Most of us at some time in our careers have worked as a one-person crew. Remember how much you could accomplish in one long day? Remember the first really good helper, the one who read your mind and did what you wanted him to do? You increased your production when you added the helper, but you did not double it.

Sidewalk snow removal (and snowplowing too, for that matter) is a series of solo, one-person tasks. Unlike construction crews, sidewalk snow removal crews do not handle heavy or awkward materials or equipment requiring more than one person to improve efficiency. This lack of synergistic benefit on a per-task basis encourages us to think of our crews as combinations of one-person crews.

Loading heavy sheets of plywood is a good example. One person can load 30 sheets per hour by himself, but a crew of two can load 75 sheets an hour. The difference is called synergy, which means the whole is greater than the sum of its parts. By working together, the plywood loading crew actually increased the output per person from 30 sheets per hour to 37.5 sheets per hour.

What would the effect be if you added a third person to the plywood loading crew? It would be detrimental to the overall production synergy.

In sidewalk snow removal work, we do not perform activities that offer opportunity for positive synergistic effect. In fact, we have the opposite. When we increase the crew size, we lose efficiency.

For example, send one person to do a job that takes 4 hours of elapsed time (*or 4 man-hours*). Now, send two people with the same equipment, the same distance, to clear the same area, and it takes 2.4 hours of elapsed time (*4.8 man- hours*). The two-person crew did it more quickly (2.4 hours instead of 4 hours), but more time was spent in *man-hours*; therefore, it became less efficient. The inefficiency translates into less profits.

LARGE CREW THEORY

In a lot of cases, clearing a sidewalk in a little more than 2 hours, rather than half a day, is a worthwhile trade off for the inefficiency, especially if it is snowing heavily at the time the sidewalk needs to be cleared.

The important issue is to recognize that the more people we send to do the job, the faster it is completed, but it is also less efficient in total man-hours spent. Your cost is proportionate to man-hours spent, not elapsed crew time. The small two- or three-person crew may not effectively clear snow quickly enough at all sizes of customer properties.

One drawback of small crews on large properties is they cannot complete the work fast enough. They spend too much time on-site, or do not get the job done on time. One answer to that problem is increasing the crew size. You should be able to send as many as six people to one property and "knock it out," then move on to the next site.

Large crews are fun to work with. They appeal to the social side of our nature, making it easy to build enthusiasm. Large crews make the members feel safe and secure. They feel as though there are enough of "us" to get it done.

Production managers and snow removal customers like large crews because absenteeism does not cripple the production effort. Supervisors (especially non-producing supervisors) like a lot of people to look after. It makes them feel needed.

Crew members like large crews, as it is like being on a team. You don't feel the pressure to produce. They have more freedom to do the things they enjoy as long as they keep busy. Property owners/managers love big crews because, in their mind, the more people running around their property the better. In fact, they sometimes demand contractors "*get more people*" on the job and "*get it done.*"

Routing and crew sizing is important to maximize sidewalk snow removal productivity and profitability.

Crews working a specific route are often sized to fit the largest property and crews seem to grow almost by themselves. Supervisors and production managers often add one member as insurance against anything going wrong. This is a sign of mismanagement, not efficiency. Everyone likes large crews except the person responsible for profit.

In some cases he does not know that large crews (with more than three people) are the problem rather than the solution. They blame people, the pricing system or the weather for the production crisis that is reducing profits.

LARGE CREW MYTHS

Increases in man-hour efficiency are only one of the many myths about large crews. Another popular myth is that large crews ensure quality work. This was born in the belief that it takes more time to do quality work, and non-quality work is faster and saves time. Neither is true.

Quality is the result of a process that includes trained people operating the correct equipment according to a set procedure. In large crews where accountability is minimal, quality is often sacrificed.

Owners/managers like large crews on site. When you are behind schedule, the first solution is to add people. Desperate property owners (when you get into snow removal on large multi-building sites) may even dictate specific crew sizes and threaten to withhold payment if these demands are not met. In most cases this "knock it out" behavior is an attempt to correct performance problems and force the contractor back on schedule. In this situation, don't increase the crew size.

The best strategy is to bring in a separate crew and divide the property into appropriate zones. Once back on schedule, the owner/manager will become accustomed to, and accept, fewer people on the site during a snow and ice event.

The myth that large crews provide better use for supervisors is a throwback to assembly-line thinking that really does not apply to on-site sidewalk snow removal crews. The notion that one strong supervisor can supervise five people as easily as two and still keep production responsibility does not apply during a snow event.

Some supervisors try to keep the men together. Supposedly, they are easier to supervise, but in reality the "herd mentality" further reduces productivity. The large-crew supervisor must make a choice to reduce his own productivity in order to keep five men up to speed, or allow their productivity to drop to maintain individual productivity.

Most large-crew supervisors do a little bit of both and lose both productivity and quality. The combination that seems to work best is a full-time working foreman with one or perhaps two crew members trained to require very little direct supervision.

Divide large crews into smaller two- and three-man crews and teach them to function as separate work units. When large properties require more man-hours than a three-man crew can generate, divide the property into two zones and send two crews to produce the work. It will be much cheaper for the customer in the long run and the snow contractor will look much better at budget review time.

Each two- or three-person crew should have production and quality goals for the specific snow event. Even though they may be in competition on the same property, they should be evaluated on that particular snow event's performance. ❄

SNOW TERM: BLIZZARD WARNING
Issued for sustained or gusty winds of 35 mph or more, and falling or blowing snow creating visibilities at or below one-quarter mile.[3]

SOURCES

[1]USA Today Weather.

[2]National Oceanic and Atmospheric Administration, National Snow & Ice Data Center, University of Colorado, Boulder, Colo., http://nsidc.org. AccuWeather.com, www.accuweather.com, National Climatic Data Center.

[3]U.S. National Weather Service.

CHAPTER FOURTEEN: Working With Deicing Products

IN THIS CHAPTER

- ❄ Deicing material and how it works
- ❄ How, where and when to apply deicers
- ❄ Pricing ice control services

Often customers will demand more of you than simply moving snow from their parking lots, sidewalks and other walkways. In today's market, most customers – especially commercial/industrial clients – demand you also keep their facilities free of ice. Liability from falls or accidents associated with ice are significant risks for commercial clients, so they often require you to keep them free of both snow and ice.

This is a good thing. Deicing activity is one of the most profitable services you will ever encounter and it is fairly easy to incorporate the service into your plowing operations. Ice control margins around the country often exceed 75 percent gross profit. The use of deicing prod-

Removing ice from parking lots, sidewalks and other traffic areas can be a very profitable service.

Deicing materials vary widely from rock salt to liquid chemical deicers. This salt pile is being treated with a liquid additive.

ucts within the framework of the snowplowing industry is a huge topic. The topic warrants an entire book in itself.

Further, what we don't know about controlling ice buildup on pavement surfaces dramatically dwarfs what we do know. The study of ice control goes on daily, primarily led by those Departments of Transportation (DOT) that must operate in areas that receive snow and/or ice each winter season. The dollars expended on such controls runs into the billions of dollars each season.

With the public's demands upon us to be able to transgress paved surfaces safely, and the ultimate demand for bare pavement, we are all learning what is needed to adequately fulfill these desires. Private snow management contractors have been driven to become more knowledgeable about how to control ice on paved surfaces.

As private snow management contractors become more knowledgeable about proper ways to manage their business, they have come to realize just throwing down more salt is not profitable, not environmentally friendly and just plain bad business strategy.

DEICING MATERIALS

For decades rock salt was the deicer of choice. Supply was plentiful, the product was effective at moderately cold temperatures and salt was a very inexpensive product to secure. The harsh winter of 1992 led to salt shortages in the northeastern portion of the United States as demand far exceeded supply. Pricing structures rose in response to the demand. Municipalities and states confiscated available salt supplies, which further aggravated the situation.

Alternative deicing products were developed in the aftermath of this salt shortage and much experimentation took place with these alternative products. Additionally, scientific methods were introduced into research studies and the push was on to develop ways to safeguard the public even in the face of a shortage of available product. The DOTs led the way in such research of methodology.

Private and public companies poured millions into development of a number of alternative deicing agents. Chlorides and chloride mixtures were formulated and tested.

One company even went so far as to test (to some degree of success) the use of distilled corn solids as a way to treat slippery pavement conditions.

The latter part of the 1990s gave North America relatively mild winters and salt supplies proved adequate for the most part. However, as private plowing contractors became more sophisticated in their own methods of servicing customers, the demand for alternative products continued.

The winter of 2000-2001 saw a radical shift in the weather patterns over North America that returned us to "old-fashioned winters." The following winter season saw contractors, municipalities, states and other public entities substantially increase their requested allotments of rock salt, which created an artificial salt shortage. This drove up pricing again.

Given the long-range (eight - 10 years) forecast that weather patterns will generally stay much the same as the 2000-2001 winter season, combined with the renewed environmental interest in how we control slippery conditions, the threat of not being able to properly service public and private customers caused renewed interest in alternative methods of controlling ice buildup on paved surfaces.

The occasional shortage of salt, combined with the desire of DOTs and snow contractors to find faster and more efficient ice-melting materials has forced creation of several relatively new technologies in deicing. These include development and use of calcium and magnesium chloride, calcium magnesium acetate and a whole host of "environmentally sensitive" products.

Manufacturers have also developed liquid chemicals, which can be applied prior to a storm, to prevent ice from forming on pavement — called anti-icing.

Techniques and products have also been developed whereby bulk rock salt can be

NOTABLE SNOWSTORM: January 1996: The "Blizzard of '96" was responsible for 187 deaths and brought much of the eastern United States to a complete halt. Schools, offices and airports were closed for several days in some areas as roads were impassable. Compounding problems, two subsequent storms blasted the same areas within the following week-and-a-half, causing $3 billion of damage.[1]

ICE I.Q.

It pays to analyze your ice control strategies.

Snow contractors often rely on tried and true methods like plowing to solve their customers' ice problems. But with liability and material costs on the line, giving their ice control strategies a second look may pay off for contractors.

That's not to say that tried and true methods don't work well. But don't be complacent, warns Dale Keep, a respected consultant and owner, Snow & Ice Technologies, LLC, Walla Walla, Wash.

"No matter what product you use and whatever results you get, ask 'Why?'" Understand the results, and why they happened. "There is always something to be learned," Keep said.

This kind of analysis can improve your knowledge of ice control strategies and your success level in a variety of situations. "Every tool has its place. You should match the product to the area you will be working in. Set out your unique guidelines and then find the best control strategy that fits that criteria."

ICE CONTROL OPTIONS. Today, a variety of ice control products are available:

- **Sodium chloride:** By far the most common ice-melting product because of its low purchase price. The commonly used effective melting temperature is 20 degrees F and warmer at a 23 percent concentration. The generally used eutectic temperature is -6 degrees F.

- **Calcium chloride/Magnesium chloride:** Both are exothermic, which simply means they release heat as they dissolve, which helps melt ice quickly. Effective melting temperature of these two products depends on the concentration being used. Calcium chloride at 30 percent concentration has a eutectic temperature of -52 degrees F, while magnesium chloride at 30 percent has a freeze point of 3 degrees F, with the eutectic temperature being -27 degrees F at 22 percent.

- **Ethylene glycol:** A liquid deicer, commonly mixed with liquid urea and applied using bulk sprayers and tanker trucks. Applied primarily at airports.

- **Potassium acetate:** Biodegradable liquid deicer primarily used for airports, but beginning to be used in other markets such as highways and parking garages. Because potassium is corrosive, it is often mixed with a corrosion inhibitor. Eutectic temperature is approximately -76 degrees F at 50 percent concentration, so it can effectively be used in colder temperatures.

- **Calcium magnesium acetate (CMA):** Developed as an environmentally friendly alternative to road salt, it is generally nondamaging to vegetation and concrete. However, the purchase price of CMA is roughly 30 times more expensive than rock salt. Effective melting temperature is approximately 20 degrees F. Dry CMA is typically used to produce a 25 percent brine.

Contractors should experiment with different deicing application strategies and chemicals to determine the best product for each situation.

TIPS AND TRICKS. While the above materials are considered the most common, there are many beneficial variations.

Paul Schipper, president, Evergreen Lawn Care, Grand Rapids, Mich., said his company primarily uses sodium chloride on parking lots and switches to calcium chloride on pedestrian areas.

John Parker, owner, New York Snow Pros, Poughkeepsie, N.Y., said he sees more and more contractors experimenting with liquid and non-traditional ice control products.

"Liquid products have not been real popular with private contractors. But that mindset is changing as of late," Parker said.

Though Parker's experience with liquids has made him a firm believer in these newer technologies, he cautions that they require added skills on the part of the contractor. "Chemicals are definitely the next step in efficient, advanced ice control, but they should not be looked at as something that can just be sprayed out and down the road you go. There is a definite educational process that needs to be followed. Liquids can backfire if you don't know what you are doing," Parker says.

Ultimately, the best strategy is to learn as much as possible about ice control products and apply them in the best-suited situation. "It comes down to looking at meeting a high level of service, with a minimal amount of chemical," Keep says.

PREVENTING DAMAGE. One way to reduce expense and deliver high-level service is to prevent damage associated with chemical ice control efforts. Most damage results not from the

Snow contractors should be careful to prevent damage to vegetation caused by deicing materials.

product itself, but quite often from misapplication or improper use. "Most products are only a problem when misused," Keep says. The most common problem is applying too much material or applying it in the wrong area.

Stephen Hewgill, national sales manager, Pursell Vigoro Canada, Inc., Tillsonburg, Ontario, says damage to vegetation usually results from shoveling snow that is laden with ice melter onto vegetation. Most vegetative damage is usually not evident until spring. To prevent vegetation damage:

- Apply product at label rates;
- Disperse snow/slush over a wide area;
- Apply ice melter early to reduce the amount needed;
- In areas that will be exposed to ice melters, install tolerant plant species.

In addition to damaging vegetation, concrete damage caused by ice melters can be costly for snow contractors. According to Hewgill, most concrete damage is not the result of chemical attack, but by the effects of thawing and freezing. Moisture seeps into cracks in the concrete and, as it changes to ice, expands and puts pressure on the concrete, causing weaker areas to crack or deteriorate. Use of ice control products increases the frequency of this process. To prevent damage to concrete:

- Apply ice melter at recommended rate;
- Remove slush and water from melting ice and snow before refreezing occurs;
- Avoid use on non-air-entrained concrete or other weak surfaces;
- Avoid use on new concrete, concrete with exposed aggregate, brick, masonry, stone or unsealed asphalt, all of which are particularly vulnerable to damage.

Using deicing material can help snow contractors reduce liability for their customers (through preventing slip and fall accidents). One study at a Buffalo, N.Y. hospital revealed that 27 percent of pedestrian-related injuries in the winter were due to icy surfaces. Contractors, too, need to protect their own liability by ensuring they have investigated a variety of ice control strategies and have the expertise to apply them in the right situation. ❄

Reprinted with permission from Snow Business *magazine.*

treated with a liquid substance that lowers the temperatures at which salt is effective at fighting ice, while at the same time reducing the corrosive nature of salt. These non-traditional materials and techniques are gaining popularity and certainly worthy of investigation.

However, by far the most common deicing technique among snow contractors is still the use of rock salt applied through pickup-truck mounted spreaders or through the use of V-box spreaders mounted on dump trucks. As a result of this predominance, most of the discussion below will be dedicated to traditional applications of deicing rock salt.

WHY DOES SALT MELT?

Rock salt, in and of itself, will not melt anything. This is why we can store rock salt in quantity. It doesn't melt by itself. In fact, it doesn't actually melt anything. The rock salt changes into liquid form when provided with a catalyst to make this happen.

The catalyst is a source of moisture. In a lot of cases, this moisture arrives in the form of snow or the ice itself. Once moisture is introduced into the equation, salt "goes to solution," meaning it dissolves into a liquid brine (23 percent salt).

This brine spreads under the ice material and breaks the bond between the ice buildup and the pavement surface. This is how ice is separated from the road, or parking lot surface. The ice can then easily be removed from the pavement by use of plowing equipment.

Because of the chemical properties of the brine, it also lowers the temperature at which the liquid will freeze by lowering the vapor pressure of the solution and thus melting the surrounding ice. With a lower vapor pressure in the solution, the solid ice wants to "go into solution" (i.e., become a liquid) and as a result melt the ice.

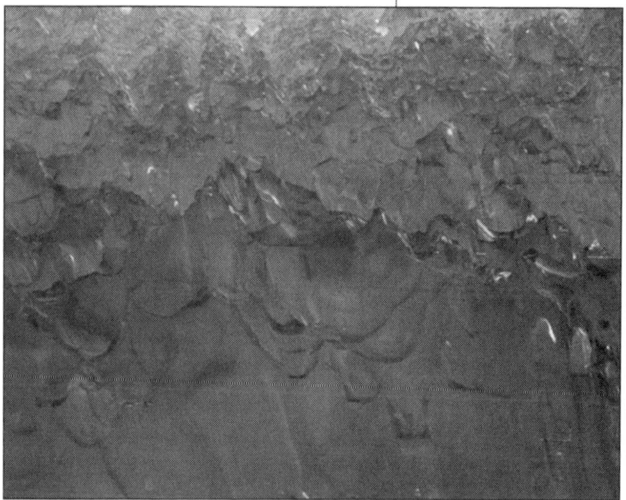

Deicing material melts ice by creating a liquid brine solution, which is up to 23 percent salt.

Heat is also obviously required to melt ice, and salt is effective at drawing heat from the air and pavement, thus making it an effective ice-melting agent.

APPLICATION STRATEGIES

Ice control on parking lots requires some capital investment.

While a truck is needed, it does not necessarily need to be a truck outfitted with a plow.

However, often the outgrowth into deicing service begins with the acquisition of a tailgate-mounted, electrically driven spreader that holds from 700 to 900 pounds of deicing material. These units are not terribly expensive and allow the small operator to enter the deicing business with minimal capital investment – usually less than $1,200.

The downside to this method of deicing parking lots and driveways is that the deicing material must be manually loaded into the hopper, thus requiring someone to be working outside in the elements several times during a typical evening plowing.

Generally, the material is loaded onto the truck in 50- or 80-pound bags, manually opened and dumped into the hopper. While inefficient from a labor standpoint, margins much higher than plowing are still generated through this method of deicing.

The next step in the evolutionary process in deicing is to purchase and mount a slide-in, V-box spreader. These can be hydraulically or electrically driven units that hold from 1.5 to 7 tons of material. Some manufacturers make units that fit nicely on a ¾-ton pickup truck; however, care must be taken not to overload the hopper. The additional weight can cause problems with the truck's suspension, causing great damage to the undercarriage and axles of the vehicle.

Once the spreader is mounted on the truck, whether on the tailgate or as a V-box, the spreader filled with salt, and the mechanism turned on, spreading the salt is as simple as driving the truck over the pavement.

Most spreaders are equipped with mechanisms that allow the operator to control the amount of salt to be applied. And once the operator determines how wide the path of salt applied by the spreader, he or she can determine how many passes are required to coat the entire lot with salt or another deicer.

Anti-Icing. Deicing is the most common means of "removing" ice from a customer's facility or residence. Deicing is the application of an ice-melting material after ice buildup has already formed on the pavement, sidewalk or other traffic area.

However, you can apply salt (or any of the alternative deicing products) prior to the snow or ice event taking place. This technique is called anti-icing and it has grown in popularity, especially among contractors using liquid deicing materials.

Salt, sitting on the paved surface, will not do anything unless moisture is introduced and comes in contact with the granular rock salt. Once it starts to snow, the moisture causes the salt to dissolve into solution. The resulting salt brine prevents a bond from forming between the ice/snow and the pavement surface. Since no bonding takes place, once plowing operations commence the snow or slush will be easily removed leaving a cleaner surface than if you plow the site after the snow/ice has bonded to the pavement.

The really nice thing is that you can achieve this result by using only one-third the amount of product required for traditional deicing. If it does not continue to snow after completion of plowing operations, there is often no need to reapply salt to the cleared surface.

If an additional application of salt is required, desired results can be achieved with considerably less material than you would have needed had you not been proactive.

All things considered, the astute contractor can actually use half the normal amount of salt by pre-salting. Most contractors who pre-salt also make a very light application of salt after the plowing has been completed.

How Much Is Enough? When discussion occurs about the use of rock salt (sodium chloride) and its distribution to the pavement surface, there are always arguments as to how much product needs to be applied to the pavement surface in order to achieve desired results.

Don Walker from the University of Wisconsin (one of the leading authorities on deicing in the country today) states that 200 pounds of rock salt on 1 acre of surface (evenly distributed) is adequate to reduce light to medium buildup of ice on pavement surface to liquid form.

DOT studies indicate that in a light-icing situation, 200-250 pounds of rock salt per acre is all that is required to reduce a light accumulation of ice to water (at approximately 28 degrees F air temperature). The melting process will take 45 to 60 minutes to complete. A heavy accumulation of ice may require as much as 350 pounds of rock salt per acre. To most readers this will seem absurdly low. However, thes low application levels are attainable.

Contractors apply salt by the half-ton and full-ton to an acre of parking surface. But more often than not, contractors apply more salt than is necessary, thus cutting into profits, leading to an overcharge of customers and doing more harm to the environment than is necessary.

PRICING AND PROFITABILITY

Much the same as snowplowing, deicing services can be priced several ways. Per-pound and per-ton methods are the most common deicing pricing methods. While easy to administer and track (for billing purposes), such pricing usually lends itself to over-use of salt and other deicing materials.

Often, the contractor doesn't even realize this abuse is happening and has no incentive to learn the correct methods of application. While this statement may offend some who have been in the business for years, the facts show it is true. Unfortunately, some of this abuse is customer-driven and not necessarily the fault of the contractor.

Customers often ask for pricing by the ton or by the pound. It is easy for the customer to ascertain (in his or her mind) what the best price is for what is required. What customers don't realize is there is really no viable way to accurately track how much product is actually applied at the time services are rendered. Unless the contractor has thousands of dollars to spend on expensive monitoring equipment used at the spreader, most just guess as to how much product has been applied.

Unfortunately, some contractors use this fact to their advantage by artificially inflating the quantity of deicing product applied so as to ensure all the salt product used is accounted for in the billing cycle.

Additionally, there is no incentive for the contractor to establish good deicing practices using the physiological facts available to him. The customer is paying by the ton, the thinking goes, so use as much as possible to achieve the desired results.

Another way the customer encourages this abuse is his perception of what takes place during the deicing process. The general public is impatient today. They want immediate results. Add to this the apparent ignorance on the part of the customer as to how the deicing process takes place, and it is easy to see why there is a high probability of abuse and over-use of deicing materials.

While this over application does speed up the melting process, it is a huge waste of product. Salt will only stay in "solution" up to 23 percent (by weight) of salt in liquid. More than that and the salt will solidify and settle to the bottom of the puddle. This is why parking lots get a white tinge to them the day after the storm has passed. This white tinge is actually rock salt that has solidified and settled to the surface of the pavement area.

Two factors also lead to overuse of salt product. One is customer-driven. The customer wants the ice gone immediately. And, if he steps onto the property and sees that the ice is not gone, he at least wants to see some material on the ice to prove you have been there. Feeling crunchy salt under his feet makes him feel comfortable, even if most of it is wasted product.

Keep in mind that rock salt will work at temperatures slightly below 0 degrees F. Unfortunately, the process slows to a point where you may wait all day for the melting process to be completed. In most cases this is unacceptable. Applying more product speeds up the process and gives the customer a feeling of security, but just wastes material.

Remember that perception is reality to most observers, we tend to overreact by applying more material. And, if the customer pays by the ton, then the customer is paying for the privilege of feeling good.

The other factor that leads to over application of rock salt is employee-driven. Most employees want to do a good job. They don't get up and go to work with the specific intent of doing a poor job. While it might appear this is the case, human behavioral studies indicate most people really want to do a good job.

When the salt truck operator leaves a site, he wants to know he has done well. Knowing that the melting process takes as long as an hour after application, it is very difficult for the operator to achieve a level of satisfaction by leaving the site knowing there is still ice or snow on that site. Thus, the employee needs the reinforcement that the process is happening and that he has done

One suggestion is to route drivers in a star pattern, so they can drive by accounts twice to check the effectiveness of their deicing application.

FOILING THE FREEZE

Using liquid deicers in private snow removal operations.

It's understandable to be apprehensive about spraying liquid on a frozen parking lot. After all, creating a hockey rink out of a customer's facility isn't the best strategy if you wish to succeed in the snow and ice removal business.

Though it seems counterintuitive, application of liquid materials to fight ice might be one of the biggest technical innovations affecting the private snow and ice removal business since invention of the snowplow. "There is a lot happening with chemicals," said Don Walker, professor emeritus, University of Wisconsin — Madison. A widely recognized expert on deicing chemicals and abrasives, Walker spoke on the subject in Denver during the 2001 Snow & Ice Symposium, an annual trade show and convention hosted by the Snow & Ice Management Association.

Liquid deicers have been around since the late 1960s when they were first used as an agent to wet granular salt before it was spread. And though it has taken nearly 30 years, liquid deicing has begun catching on with private snow and ice professionals. Today, there are not many snow pros who haven't at least considered incorporating liquid into their service mix.

Robert Wilton, president, Clintar Groundskeeping Service, Markham, Ontario, began pre-wetting salt with liquid deicer last year and sees it as a transition to a time when he will use liquid-only applications. "We didn't have the leap of faith to give up the granular yet," Wilton said.

"Everybody is thinking about it, but nobody has bit the bullet to do it," said Brian Akehurst, grounds maintenance operations manager, Akehurst Landscape Service Inc., Joppa, Md. "I think within the next couple years, we will be doing liquid applications."

Currently, Akehurst Landscape pre-treats bulk piles of deicing salt with a liquid deicer, but concerns about temperature have kept it from using liquid only. "Usually, when you use a liquid deicer, the ground temperature needs to be below 32 degrees to work effectively," Akehurst said. "If not, it can create major problems and create slippery pavement. You have to know your stuff before you go out and do it, because you can cause more problems."

Sam Granados, owner, Integrated Striping Systems Inc., Golden, Colo., has used liquid deicing since 1996.

Granados credited liquids with transforming his company into a full-service operation and allowing him to pursue large accounts. "Once we got involved with the liquids and we used them as part of our operation, we were able to offer a full level of service, everything from pre-treating to deicing to clearing ice buildup," Granados said.

Anti-icing, the application of ice control material before a storm, can make snow removal easier and help contractors deliver bare pavement.

There's no question that liquid deicers, on balance, melt faster and last longer than traditional salt applications. But they are more expensive, both in material and equipment costs and in the additional training it takes to become competent in using them. "It's not like you can go out, spray it, and call it good," Granados said.

For most companies the decision comes down to economics. "You have to look at the costs and where you are putting it down," Walker said. "Are the faster action and corrosion benefits worth the cost? I can't answer that for you. But that is a trade off you have to look at."

There are two basic techniques to using liquid deicers — pre-wetting and anti-icing — and the many liquid products available.

PRE-WETTING. Pre-wetting involves taking a liquid product and spraying it on a dry product, usually rock salt. Pre-wet materials stick to the pavement better and begin melting the ice faster.

Rock salt needs to form a brine solution (23 percent solution) to melt ice. Pre-wetting, and the added moisture it brings, speeds the brine formation process. "To me that is what you are buying when you get into pre-wetting, the ability to act faster and stick to the surface," Walker said.

Akehurst has witnessed firsthand the time savings associated with pre-wetting. The company had a parking lot where one portion was spread with treated salt, the other with dry rock salt. "When we were leaving the site the treated salt area was already black, it had already melted," Akehurst said. "Where we put down the regular salt it was still working and it had not melted yet. So I am definitely sold on the treated salt."

But there's a price for that quick action. Pre-treat rigs can range from $1,150 for a pickup truck-mounted system to $2,500 or more. Chemicals that can be used to pre-wet salt range from salt brine to chlorides to organic-based products.

"The theory is that you use substantially less salt," Wilton said. "So our cost should not be any higher, but in reality we still have to train all the operators that more isn't better," Wilton said.

However, Wilton has passed the extra cost on to the consumer in many cases, because his Fortune 500-type clientele see the reduction of salt use as a benefit to their overall corporate environmental strategy.

Akehurst added, "What we decided to do, and I'm not recommending this for everybody, but we told our clients that this year is our test year, so we are not going to charge more. We are going to see how it goes, prove the point and then next year we will talk about price."

Indeed, a fair amount of customer education is involved in using liquid deicers. Akehurst recently received a call from a skeptical customer who wondered why another snow company was out treating the lot next door, while she hadn't seen any Akehurst trucks that day. "So we had to explain how we didn't need to come out because (the pre-wet salt) kept (the pavement) from refreezing," he said.

ANTI-ICING. Less common than pre-wetting is the application of liquid deicers before a snowstorm to melt the first layer of precipitation and prevent the buildup of snow pack, that icy layer of snow that bonds with the pavement, making removal frustrating and time consuming. "If you have areas where you want to provide a very high level of service, where you want to keep it bare and get the pavement bare as soon as possible after the storm, anti-icing may have some benefits for you," Walker said.

The beauty of anti-icing is that once the liquid is applied to the pavement, it can provide residual benefits for many days. So, if a predicted storm doesn't hit or hits late, the liquid will still be on the pavement waiting to do its ice-melting job. The chemical will be washed away by rain, however.

Temperature can also be an important factor. Using these chemicals when it is too warm can have disastrous effects by creating slippery conditions. "These chemicals really react to temperature, not only the amount (of material) you need, but the time it takes them to work," Walker added. As a result, using liquid for anti-icing requires a well-trained and sophisticated operator.

"I suggest that if you get into anti-icing, you need to spend some money on weather forecasting," Walker said. "You have to know what is coming. If your weather forecasting is the 6 p.m. news, then you don't belong in anti-icing," Walker said. In addition to professional weather forecasting services, a handheld thermal scanner, which can read pavement temperature, is recommended.

MATERIAL	EUTECTIC TEMP	EFFECTIVE TEMP	COST	LIQUID APPLICATION	CHARACTERISTICS
Salt ($NaCL$)	-6F	15-20F	\$25-\$45 per ton	Can be made into liquid by creating 23% solution known as salt brine	Requires heat and water to dissolve salt; Melting takes time; Corrosive
Magnesium Chloride ($MgCL_2$)	-28F	0F	50-70 cents per gallon	Pre-wet at 6-10 gallons per ton of salt, liquid (22-30% concentration) applied at 13-33 gallons per acre	Melts ice quickly; Special handling required; More costly than salt; Corrosive
Calcium Chloride ($CaCl_2$)	-60F	-20F	30-65 cents per gallon	Pre-wet (32% solution) applied to salt at 8-10 gallons per yard of salt); Liquid only (32-42% solution)	Melts ice quickly; Special handling required; Adds 10-20% to cost of salt when pre-wetting; Corrosive; May leave pavement wet
Potassium Acetate	-76F	-15F	N/A	Pre-wet at 1.25 gallons per 100 lbs. of solid deicer, 1/2 to 3 gallons per 1,000 square feet	Low corrosion; Bio-degradable; Low toxicity to fish and mammals; Safe for vegetation; Does not track; Polyethylene containers for storage
Calcium Magnesium Acetate (CMA)	-18F	20F	\$1.25-\$2.10 per gallon	19 to 40 gallons per acre	Very low corrosion; Low environmental impact; Can reduce corrosion of salt when used as pre-wet; Costly
All Clear™, Caliber™, First Down™, Ice Ban™, Magic-0™, NC-1000, NC-2000, NC-3000 (All these products are quite different, and should be investigated individually.)	-30 to -85F	Varies	65 cents to \$1.20 per gallon	Application rate varies	Various deicer blends with agricultural bi-products from processed corn, sugar, distilling or brewing; Non-corrosive; Low environmental impact; Reduces corrosion when mixed with chlorides; Costly

In addition to its trickiness, liquid anti-icing can be expensive. A pickup-mounted liquid application system can cost \$3,000 to \$8,000 depending on the gallon capacity. A high-capacity tank for a 5- or 10-ton dump truck can cost \$10,000 or more.

Despite the expense, it is clear more private snow and ice professionals are exploring how liquid deicers might fit into their businesses. The technology is not a magic bullet, experts caution, but it can be a very useful tool in many situations. ❄

Article reprinted with permission from Snow Business *magazine.*

By offering per-occurence pricing for deicing services, snow removal professionals can reduce the overall cost to the customer, while increasing profits.

a good job. He too wants to see results. This is normal.

Wanting to see results, the employee might be tempted to apply more material in an effort to speed up the ice-melting process. Here again, education and

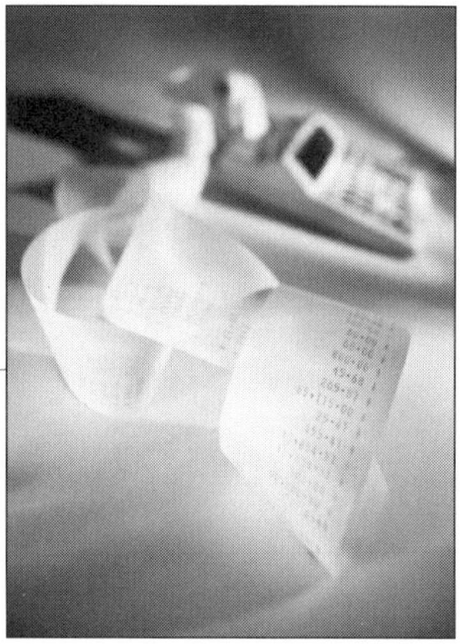

knowledge are key factors in successful operation of an ice control program. Even if you properly educate the operator as to how much material is required in order to achieve acceptable results, it will go against the grain of human nature.

Assuming the operator has been properly educated as to how much material is required to achieve desired results – giving him the visual confirmation he craves can be done.

One suggestion is to route the driver in a star (or criss-cross) geographical pattern. This forces the operator to go by sites that he has already treated.

It is inevitable that he will visually inspect the previously treated site while driving by. This will assist in the positive reinforcement needed in order to feel he performed satisfactorily. Granted, this may not be possible in all instances, but even some positive reinforcement is better than none.

Interestingly enough, this same premise works well with sidewalk crews who are required to melt ice on sidewalk areas after clearing snow. The minor inefficiencies created by not designing routes in a straight line will be offset by the dramatic reduction in material needed on a site, as well as the additional time spent over applying material on that site as well.

"Ah," says the profit-minded reader. "I'm charging per-pound for the material and per-hour for my labor and you have just reduced my necessary labor output, as well as the amount of material I can charge my customer for using." This is true.

However, it should become increasingly more apparent to the profit-minded contractor that margins can be increased dramatically if the customer was charged per-occurrence.

For argument's sake, let us assume you are charging per-pound for material and per-hour for labor. It should be easy enough to compute what the customer has paid (generally speaking) per visit to the site to perform deicing services. Knowing full well you are over applying material, and that you probably can reduce the material cost by a significant amount through education and proper training, why not figure out what the customer has been paying (on average) each time you visit the site? Then go to the customer and lay it out in detail showing him what he pays you.

During the meeting offer a per-occurrence pricing strategy that lowers the customers' total per-visit cost by 15 or 20 percent. In this arrangement, everybody wins.

The customer pays less and admires you for reducing his cost. You increase margins by reducing the material required (and reducing the necessary labor on the site).

Now, the caveat to this is education. Operators need to be properly routed. They also need to be properly educated as to how much product is actually necessary in order to achieve the desired results. And, you have to monitor what they are doing in order to be certain everyone is functioning as they should and applying the appropriate amount.

Some snow contractors insist on using a mix of sand and salt for deicing, despite sand's questionable effectiveness.

Pre-salting is the application of deicing material before a storm begins.

Alternative deicing materials and equipment, such as the liquid deicing sprayer pictured above, are becoming more common.

This is known as managing. Once you begin managing your business, you'll see increased profits and happier customers.

Let's take this a step further. If you are charging a customer a flat seasonal fee, which includes salt applications, there are other things you can do to increase your margins and service the customer better.

Under the seasonal price strategy, one way contractors have increased margins is to pre-salt parking lots. Often referred to as anti-icing, this is one way to dramatically reduce material costs normally associated with ice control. It also gives the customer a better result.

Again, we can hear the traditional plowing contractor thinking that this is a hard sell to his customer. And, for those who charge per-occurrence or per-pound, this is true. It is very difficult (in the private sector) to convince property managers and owners this is a wise use of their money. Getting a property owner to accept the fact that you are going to salt his or her lot *before* the storm, and (possibly) when the sun may still be shining is difficult at best.

However, on those accounts where customers pay one fee for the entire season, it is in the contractor's best interest to utilize the most efficient and cost-effective methods available. In this manner, customer service levels are actually heightened while at the same time increasing margins for the contractor.

SAND/SALT

While we are on the subject of rock salt, let's take a look at the use (or rather misuse) of sand as a traction agent. Sand as a traction material has a very short effective life span. This product is erroneously used in markets under the guise of being an adequate way to control sliding on ice by vehicular traffic. It is not. The effectiveness of sand on roadway surfaces as a traction control product is poor at best.

Studies by the Iowa and Idaho DOTs, have shown that sand is effective for traction by as little as five and as many as

10 vehicles before losing all traction ability.

However, in certain markets, contractors and customers alike insist on using predominantly sand mixtures for deicing and traction control on parking lot surfaces in spite of the numerous studies to the contrary. The progressive contractor who knows the facts will shy away from the use of sand for deicing in lieu of the materials that are better suited for deicing.

Another reason that sand is not a good material for use as a deicing and/or traction agent is the inevitable cleanup required at some point in the future. Sand clogs storm sewers and collects along curbs. Picking up sand material in spring can be costly to the customer. While a source of revenue for some, it is not in the customer's best interest.

ALTERNATIVE DEICING MATERIALS

As noted above, salt is still clearly the most common deicing material in the market. However, alternative deicing materials have gained credibility among contractors in recent years. The liquid products, though more expensive than salt (both in material cost and the cost to apply them) can be highly effective if used properly.

What's more, because they melt faster and have a longer residual life than salt, these new deicing chemicals can boost the efficiency of a contractor's ice-melting operations. Further, a few organic-based deicers are available, which can be used in various environmentally sensitive situations.

Unfortunately, lack of knowledge impedes the use of alternatives to salt more than a lack of the products' performance. All of the liquid materials have very specific uses, temperature ranges and usage requirements. If not used as directed, performance levels fall and satisfaction levels drop.

There are not nearly enough pages in this book to adequately explain all of the different products available to the market. What's more, the technology is changing so rapidly that any lengthy discussion would be out of date shortly. ❄

SOURCES

[1]National Oceanic and Atmospheric Administration, National Snow & Ice Data Center, University of Colorado, Boulder, Colo., http://nsidc.org. AccuWeather.com, www.accuweather.com, National Climatic Data Center.

[2]National Snow & Ice Data Center, University of Colorado.

[3]USA Today.

Chemicals and Abrasives, presented by Don Walker, University of Wisconsin, Madison, Wis.

CHAPTER FIFTEEN: Managing A Snow Event

IN THIS CHAPTER

- ❄ Preparing for the storm
- ❄ Fighting a storm
- ❄ Dealing with the extreme snow event

The most important part of any snow business plan is to be ready, willing and able to go to work if, and when, it snows. Implementation of the overall plan is crucial to success. What good is it to have all the business aspects covered, all the equipment ready and a plan in place, only to fail when the snow comes. Let's take a step-by-step approach to how contractors approach an actual event.

Most snow contractors routinely monitor the weather during the snow season.

BEFORE THE STORM

1. Weather Monitoring. The snow removal process starts with the advanced prediction of a snow event. The local forecasters usually give an advanced warning at least three days prior to the date of the event. Keeping in mind that television forecasters are predicting for the masses, the astute contractor will not use this general forecast as gospel. However, it is an advanced warning of what might be ahead. Many contractors religiously monitor *The Weather Channel* during the winter season and some even subscribe to satellite weather services, which provide on-call or Internet-based, 24-hour weather prognostication and surveillance.

2. Review Equipment. Three days prior to the event, a review of the available snow removal equipment is advisable. The equipment should already have been tested and made ready for winter, however, one last look will assure that no surprises are

As the snow and ice event gets closer, appropriate telephone communication with subcontractors is vital.

in store when the snow falls. If you have gone several weeks since the last storm, ensuring that the equipment is in working order can save considerable headaches when the snow is flying.

After checking the equipment you are using, it is also advisable to contact your subcontractors. Sometimes these fellows are not paying as much attention to the weather as you are, and a short phone call to give them a heads-up can save considerable confusion when you call them out at midnight a few days down the road.

3. Working The Phones. As the actual time for the snow event gets closer, an additional phone call to subs and employees the day of (or the day before) the predicted event will tell you if you have any surprises coming. A sick employee can possibly be replaced with a back-up person if you are aware of the illness. The same goes for the subcontractor who might be on vacation, out of town or just plain not ready to go. Often, especially with employees, an additional phone call during the daytime hours (assuming a nighttime storm) will alert the employees you will be calling. The "I-didn't hear-the-phone" excuse can be eliminated if they know in advance the call is coming. After the calls are complete, then supply the dispatcher with the list of who is not available as this can save considerable time at the beginning of the plowing event.

Usually, the person who will make the decision to "go" will not get much sleep the night prior to the storm's arrival. Constant attention to the local forecasts, a satellite feed of current radar and what's actually happening outside will take up most of this person's time.

DURING THE STORM

1. Monitor Your Customers. When it starts snowing, the decision-maker will often go out and drive around to see the conditions in his market area. He may gain input from others who are on alert to watch the conditions. In areas where weather can be different within the same general market area, input is usually necessary to know how many crews to call out to work. When the trigger depth is reached in a specific geographic service area, the decision to plow is made. Phone calls to the dispatcher and the area supervisors should be made at this time, and usually from the decision-maker's vehicle via cell phone or CB radio. The dispatch team arrives at the office within an hour. The area supervisors are calling out their crews. Crew leaders then call out their operators. Ideally, the entire call-out process is completed within an hour of the decision to "go."

2. Dispatching. Crews should have preordained routes to cover. These routes are usually geographically clustered to eliminate undo travel between sites. Proper and efficient routing is essential to the successful completion of the plowing operation. Crews then proceed to those route areas and begin service. The dispatcher will begin communicating with area supervisors to determine where the company is "short" on drivers, equipment operators or subcontractors. The dispatcher will communicate with customers as needed. Continual communication between the dispatcher and area supervisors will keep everyone informed of the conditions in the field, the progress of plowing operations and any unusual events that might occur (accidents, breakdowns, changes in storm intensity). The dispatcher will be recording all of these benchmarks in chronological order either via dictaphone or noting everything on paper or on a computer . Often, the area supervisors carry dictaphone-type equipment to record their own observations. Such tapes are then transcribed after the event and placed in a file for future reference, just in case. Remember, this kind of proper record keeping is essential to a snowplow operation and allows for proper billing, customer communication and liability protection in the event of an accident.

Assuming that plowing operations have started about midnight, all area supervisors should be reporting to dispatch at about 5 a.m., to update the office as to what is happening. Hopefully all is well and crews are proceeding in an orderly fashion to clear snow at all assigned locations. However, in the unlikely event that some things did not go as planned, this is a good time to alert everyone where any problems might exist and what is being done to expedite changes in schedules to accommodate the problem areas. The dispatcher has an overall view of the progress and can then assist in making decisions as to where to reroute a crew (or a plow unit)

NOTABLE SNOW-STORM: January 1998: Intense ice storm hits Maine, New Hampshire, Vermont and New York, causing more than $1.4 billion in damage and 16 deaths.[1]

A SNOW OPERATION SYSTEM

Great companies are not built on great people. They are built on great systems. Nowhere are systems more important than for contractors in climates with regular snow. Contractors who fight snow and ice storm events must have some sort of system in place to attack the storm methodically if they are to be profitable.

In most parts of the country, the total number of snow events is less than 15 annually. With such a small number of times working a snowstorm, developing a system that everyone can memorize is not possible because there just aren't enough occasions to practice how to operate in times of crises (which is exactly what most snow or ice storms tend to be). Therefore, a system should be in place so everyone knows what is to be done during any given snow event.

Assuming you have a number of pieces of equipment working a snow event, and assuming some of these are subcontractors, a system can be developed that will work for most storms. For purposes of this article, we will assume you are a medium-sized contractor using 25-plus pieces of equipment (or would like to be in this position). How should such a system operate? Your system may vary slightly from the one described below, however, this system is in place in several states and operates satisfactorily.

CALLING ALL PLOWS. Most contractors operate a reverse pyramid system for calling out plowers. One person is responsible for deciding the crew has to get rolling. Once that decision is made, calls are placed to those who are in charge of a particular geographical area. These area supervisors then call out the crew leaders who are responsible for one to four plow routes. The crew leaders call out those plowers who work under them. Area supervisors can usually handle four to five crew leaders. Crew leaders can normally handle four to five routes. Each route can consist of upward of four plows (or pieces of equipment). Plowing accounts are plowed in turn and normally done on a set schedule depending upon when the customer's property needs to be cleared.

Once you have three area supervisors, you need a dispatcher to keep track of (and have control over) everything and everybody in the field. The person who decides to start plowing also calls the dispatcher to work the radios and phones. Most plowing operations put the dispatcher in charge of operations during a snow event.

In addition to making sure subcontractors are ready to go, the dispatcher should call out sidewalk crews when needed.

Areas of responsibility are as follows: frontline plowers make certain the accounts assigned to them are plowed in a fashion consistent with what the customer wants. Crew leaders plow some accounts themselves (although not as many as the frontline plowers) and check to see that the accounts assigned to their route are plowed correctly. This might include doing some touch-up work by tucking corners, cleaning up feathers and squaring off piles. Area supervisors have a couple of accounts to plow, but most of their responsibility is to fill in the gaps when frontline plowers break down, don't come out to plow or cannot plow during an event. Additionally, the area supervisors are responsible for taking care of call-backs and special requests from dispatch. Ultimately, the customers' satisfaction is the responsibility of the area supervisor. Additionally, the area supervisor should be considered an overhead expense and not a direct expense to the plowing route's profitability.

Assuming you are paying subcontractors and plowers by the hour, the area supervisors should be responsible for collecting the hours worked during each snow event. Frontline plowers turn their hours into the crew leaders, who turn these hours (and their own) into the area supervisor. The area supervisor turns all hours for his/her people into dispatch, either in person or via phone. The area supervisor should also report whether or not all accounts have been plowed as directed and be responsible for recording who plowed what account and when (if your record keeping demands these details). If some accounts are being charged per-hour/per-piece of equipment, then the area supervisor records these times and turns them into dispatch.

It is also a good idea to have the frontline plowers call your office to report their hours worked too. This way you have a method of double checking to be certain that all hours have been recorded accurately. If there are any discrepancies, you will catch them immediately if the hours are reported twice. If your subcontractors are billing you monthly, it is often hard to go back and reconstruct plowing events that may have happened almost 30 days ago. In the event of a dispute, the area supervisors' records should prevail. After all, they are getting paid extra to handle this important reporting function.

A systems approach is crucial to snowplowing success.

If you have an assigned dispatcher, there should be regular contact between dispatch and the area supervisors. Constant updates about the status of each route and area will ensure there are no surprises when you begin to wrap up the plowing for that particular event.

Some companies require the dispatcher to check with all area supervisors at a specified time to assess what still needs to be plowed. If possible, the area supervisors should be able to communicate with each other during the snow event. This way they can stay on top of what still needs to be done. Additionally, the dispatcher can then move people around to make certain all accounts are properly cleared in a timely fashion. Normally, if a route is completed and it seems like it is time to begin sending people home, the dispatcher can make that decision depending upon the status of other routes. Having a competent dispatcher is the key to the success of any ongoing plowing operation.

DISPATCH DUTY. Once the decision has been made to mobilize for a snow event, the role of the dispatcher can be enhanced greatly. In addition to overseeing the progress of the plowing operation, the dispatcher can call out the sidewalk crews and the salting crews. These crews typically go out to work some time after the plowers begin operations.

The dispatcher can contact the sidewalk crew members and the salt truck operators. Sidewalk crews generally don't live together and must meet somewhere to begin their work. If that meeting place is your warehouse or office (and assuming that the dispatcher is located in that building), the dispatcher can record start and end times for the crews. Such meetings are also a good way for the dispatcher to know who hasn't shown up for sidewalk work. Then the dispatcher can begin the process of calling in alternate workers to fill in for the conveniently missing crew member.

Additionally, the salt truck operators normally will not have full and ready trucks at home and must be called in to begin their work. This work will include checking the truck fluids, filling the salt/sand hoppers and proceeding to their first stop. Assuming that the salt truck operators can communicate with the dispatcher, they can be dispatched to on-call customer locations when a request comes in for service. The dispatcher usually pays close attention to what has been plowed and can dispatch the salt truck or sand truck to locations that are plowed and ready for deicing product.

At the end of the snow event, and if the system is in place and operating, everyone involved will feel a sense of accomplishment. Chaos is avoided with an in-place system, and rough spots throughout the snow event can be smoothed over if a successful system is utilized.

Fatigue is a normal occurrence in the snow and ice control industry. Not having a system in place can exacerbate that fatigue and cause tempers to flare from an apparent lack of organization.

Additionally, having a system in place allows for controlled growth. Once a good system works, adding to it is easy. Adding more area supervisors becomes a matter of hiring or promoting someone. Then they just blend into the system and growth continues smoothly.

At some point you may have to add another layer of management; however, that addition then becomes a natural extension of the system already in place. And, unless you have 40-plus snow events during the winter, this system should be written down and available to all who work within that system. You cannot expect everybody to remember what is to be done if you do not have the opportunity to practice the system you have in place, and most contractors don't have the luxury of using their system regularly because of the nature of the snow business.

Great companies are not built upon great people. There simply are not enough great people to go around. Great companies are built on great systems. Once the system is in place, let it work. Then get some sleep. ❄

SNOW FACT: More than 16 million tons of deicing salt were sold in 2001.[2]

to shore up the area that may be behind schedule.

As previously mentioned, the area supervisor is responsible for keeping track of what has been plowed and what has not yet been completed. The dispatcher(s) is responsible for knowing the status of the various areas as far as completion of account plowings.

3. Salt Runs. If salt trucks are dedicated and do not have plows on them, these units usually start sometime after the plowing crews have begun operations. It rarely does any good to salt the lot prior to plowing operations. If a company does anti-icing — the application of deicing material before a storm hits — it is done hours before the start of the snow event. If the plow trucks carry deicing equipment (tailgate spreaders or V-box spreaders), deicing operations generally take place immediately after completion of the plowing operation on a particular site. Proper scheduling of the deicing operation can greatly enhance the look of the finished product.

4. Quality Control. Area supervisors and crew leaders are responsible for completion of plowing operations. They are also responsible for the quality of work their crews are doing.

A high quality job is one where the customer looks at the site after plowing operations are completed and is satisfied with what he sees. Corners are square and tight with no snow feathering out into the lot. The plow lines are straight and not wavy. Light poles don't have unsightly snow piles haphazardly left behind. Curb lines are evident and piles are tight in specified snow dump areas.

Fire hydrant locations are evident since the hydrants can easily be identified. Snow is not piled upon sidewalks or left in driveway aprons. And no snow has been piled in the street or drive lanes. A quality job is easily identified and poor jobs are not acceptable.

5. Paperwork. As the storm comes to an end and operations are coming to a close (assuming it has stopped snowing and ongoing plowing operations are not required), crew leaders are turning in hours and completed plowing data to the company's area supervisors.

Area supervisors are checking sites for quality control and may be touching up the lots to make sure all looks good to the customer when he arrives on site. Area supervisors are then turning in hours and completed data to the company's dispatcher who checks this information against his own in-house information-gathering system to ensure accuracy. Everyone knows that the job is not done until the paperwork is completed.

Subcontractors may then be calling in to double check that their hours were properly turned in and to be certain that their records match those of the company. The benefit is that everyone is aware that such a double-check system ensures all parties will be paid the correct amount at the correct time.

In the case of ongoing snowstorms, these checks and balances are necessary since operators can often get confused if plowing every day for several days in a row.

6. Clean-up Work. The next night, the crews may be called out again to perform clean-up work on sites. Packed snow gets mushy, requiring crews to clean that slush off the parking lots to keep them safe. Possibly piles have to be relocated to accommodate the needs of the customer. Often reapplication of deicing materials is required due to melt off and refreeze. Area supervisors can often make these decisions in conjunction with the dispatch team.

A word of caution: Many companies attempt to utilize speed crews for clean-up work. The speed crew is usually made up of experienced snowplowers who know the sites. It is thought that these plowers will get through the sites quickly if the snow accumulation is light or if clean-up work is all that is necessary. The experience of seasoned snowplowing companies indicates that this may not be the most effective way to address such circumstances. Those not on the speed crew are often offended that they were not utilized for the easy work. And often those on the speed crew are hard pressed to get the work done on time as undue pressure is placed on these individuals to perform very quickly. Quality is sacrificed in lieu of time constraints and the customer can become disillusioned with the finished product.

Thorough documentation during a storm is important from a number of perspectives.

Extreme snow events will fully test your abilities and company-wide systems.

Experienced plowing companies often realize that sending the entire area crew out to do the clean-up work is beneficial in many ways. Everyone on the crew(s) then learns what is expected in the clean-up process. They also get some of the easy work instead of feeling they were hired only to beat up their equipment during bad storms.

In the long run, these operators are usually grateful for the trust placed in them to do an efficient job during the clean-up operation and will work faster knowing they will get called out the next time there is an easy night coming.

7. Follow-Up. The last thing that can be done as part of the snow clearing operation is to follow-up with the customer. Customer follow-up can be the most important part of the plowing operation. A satisfied customer is much easier to keep than finding new customers.

And, if there is a problem, it is best to find out about it at the time rather than to find out at contract renewal time in the summer months when there is little you can do to rectify the customer's concerns.

THE EXTRAORDINARY SNOW EVENT

Every once in awhile a snowfall occurs that is truly exceptional for the market area. Depending on the market, this type of storm might occur only once in a decade, while other markets may experience one extra heavy storm per year. In some markets, the catastrophic event might be 6 inches of snow overnight. In other markets such an event might be defined as 3 feet of snow in a 24-hour period.

In any case, these events must be dealt with differently. Back-up equipment is often necessary, and this equipment should be lined up prior to the season. Such avenues for back-up equipment might be as simple as having the home phone number for an equipment rental outfit that can supply you with loaders if needed in an emergency.

Keep a list of plow drivers who are willing to work with you if a bad storm arises — these might be people who have small routes of their own and would be willing to work for you to gain extra income in a bad storm. Some companies have names and numbers of plowers from as far as two hours away, who can be recruited in the case of a monster snowfall.

Plans should be in place for relief to plowers and operators who are involved in an ongoing storm situation. While constant plowing during the storm is a good way to keep up with monster snowfalls, it must be kept in mind that a fatigued loader operator can do considerable damage to property if not given time to sleep. Often, plowers only need time to shower and eat in order to continue working over the long haul.

Damages occur in increasing frequency as plowers become fatigued by an ongoing storm. It does a company little good to have a subcontractor damage his own equipment and not be available to continue plowing. Thus, it is often much cheaper to let subcontractors and exhausted employees go home for a few hours, rather than to lose them for a few weeks.

Record keeping and customer communication become increasingly more important when a catastrophic storm occurs. Supervisors and dispatchers need to communicate often so that proper documentation is taking place during the storm.

Customers will call stating, "you haven't been here yet," and you need to know whether or not that is accurate. It is very difficult to explain to a customer (when he gets the invoice) that you were there two or three times providing services when he is certain you were not there. This often happens because the customer looks out the window and sees "white" but doesn't see the piles you made during a previous visit the same day.

Sometimes a simple fax to a customer can alleviate this misperception. If all incoming calls for additional service are routed to the dispatch center, and assuming the dispatcher has the information, then the customer can be informed that "we were there around 4 a.m. It is continuing to snow and we can return to re-plow if you so desire." In this way thc customer is informed that service has been provided. A follow-up fax to the customer then gives them this information in writing so there can be no confusion when the invoice arrives in the customer's office location. ❄

SNOW TERM: SNOWPACK
The total snow and ice on the ground, including both new snow and the previous snow and ice which has not melted.[3]

SOURCES

[1]National Oceanic and Atmospheric Administration, Boulder, Colo., http://nsidc.org. AccuWeather.com, www.accuweather.com, National Climatic Data Center.

[2]The Salt Institute.

[3]National Snow & Ice Data Center, University of Colorado.

INDEX

A

B

C

D

E

F

H

I

T

W

PHOTO/ILLUSTRATION CREDITS

COVER IMAGE

.... Boss Snowplow

ABOUT THE AUTHOR

Page 7 .... Paul Lorei

CHAPTER ONE

Page 17 .... Donn Angus
Page 18 .... National Oceanic & Atmospheric Administration
Page 19 .... National Oceanic & Atmospheric Administration
Page 20 .... PhotoDisc

CHAPTER TWO

Page 23 .... Darryl Leitsinger (Photograph of Dan Foley, D. Foley Landscape, Walpole, Mass.)
Page 24 (top) .... Stockbyte
Page 24 (bottom) .... Stockbyte
Page 25 .... Fisher Engineering
Page 26 .... *Snow Business*
Page 27 .... TrynEx International
Page 30 .... Art Explosion/Nova Development
Page 31 .... Steve Smith

CHAPTER THREE

Page 33 .... PhotoDisc
Page 34 .... PhotoDisc
Page 35 .... PhotoDisc
Page 37 .... Sierra Studio (Photograph of Chris Frederick, Frederick Corporation, Blue Island, Ill.)
Page 38 .... PhotoDisc
Page 39 .... Stockbyte

CHAPTER FOUR

Page 41 .... Mainscape, Indianapolis, Ind.
Page 42 .... Stockbyte
Page 44 .... Steve Smith
Page 46 .... Wayman Williams (Photograph of Chris James, Chris James Landscaping Inc., Midland Park, N.J.)
Page 49 .... Roger Mastroianni
Page 51 (top) .... Wayman Williams
Page 51 (bottom) .... John Allin

CHAPTER FIVE

Page 53 .... GIE Media Inc.
Page 54 .... Donn Angus
Page 55 .... Chris James
Page 57 .... Snow Wolf

Page 60 .... PhotoDisc

CHAPTER SIX

Page 63 .... PhotoDisc
Page 64 .... Darryl Leitsinger (Photograph of Dan Foley, D. Foley Landscape, Walpole, Mass.)
Page 65 .... Boss Snowplow
Page 67 .... Chris James
Page 68 .... Martha Muscarella
Page 71 .... PhotoDisc
Page 73 .... PhotoDisc
Page 74 .... GIE Media Inc.

CHAPTER SEVEN

Page 77 .... PhotoDisc
Page 78 .... Donn Angus
Page 79 .... PhotoDisc
Page 81 .... PhotoDisc
Page 84 .... Ariens
Page 87 .... Donn Angus

CHAPTER EIGHT

Page 89 .... Comstock
Page 90 .... Comstock
Page 92 .... Comstock
Page 96 .... John Allin

CHAPTER NINE

Page 99 .... Stockbyte
Page 100 .... Stockbyte
Page 103 .... John Allin
Page 107 .... John Allin
Page 111 .... PhotoDisc

CHAPTER TEN

Page 113 .... Steve Smith
Page 114 .... David Haskell (Photograph of Sam Granados, Integrated Striping, Golden, Colo.)
Page 118 .... GIE Media Inc.
Page 121 .... Hiniker
Page 126 .... PhotoDisc
Page 127 .... GIE Media Inc.
Page 128 .... PhotoDisc

CHAPTER ELEVEN

Page 133 .... Carol Phelps
Page 135 .... Hiniker
Page 137 (clockwise from top left) Hiniker, Meyer, David Haskell, Daniels Plow, Boss Snowplow
Page 139 .... Boss Snowplow

Page 143 (left) .... Chris James
Page 143 (right) .... Snow Wolf
Page 144 .... Western
Page 147 .... Snow Wolf
Page 148 (left) .... Sweepster
Page 148 (right) .... Simplicity Manufacturing
Page 150 .... Western

CHAPTER TWELVE

Page 153 .... Paul Lorei (Photograph of John Allin, Snow Management Group, Erie, Pa.)
Page 154 .... Sweepster
Page 155 .... GIE Media Inc.
Page 156 .... Karen Angus
Page 159 .... Snow & Ice Management Association

CHAPTER THIRTEEN

Page 161 .... Toro
Page 162 .... John Deere
Page 166 .... GIE Media Inc.

CHAPTER FOURTEEN

Page 169 .... Art Explosion/Nova Development
Page 170 .... Chris James
Page 173 .... Hiniker
Page 174 .... PhotoDisc
Page 175 .... PhotoDisc
Page 176 .... TrynEx International
Page 179 .... Western
Page 181 .... John Allin
Page 184 .... Stockbyte
Page 185 (top) .... C.U.E. Inc.
Page 185 (bottom) .... TrynEx International
Page 186 .... David Haskell

CHAPTER FIFTEEN

Page 189 .... Wayman Williams (Photograph of Chris James, Chris James Landscaping Inc., Midland Park, N.J.)
Page 190 .... Sierra Studio (Photograph of Chris Frederick, Frederick Corporation, Blue Island, Ill.)
Page 193 .... Ariens
Page 194 .... Boss Snowplow
Page 197 .... Wayman Williams (Photograph of Chris James, Chris James Landscaping Inc., Midland Park, N.J.)
Page 198 .... John Allin

NOTES

NOTES

NOTES

SNOW REMOVAL BUSINESS FORMS ON CD-ROM

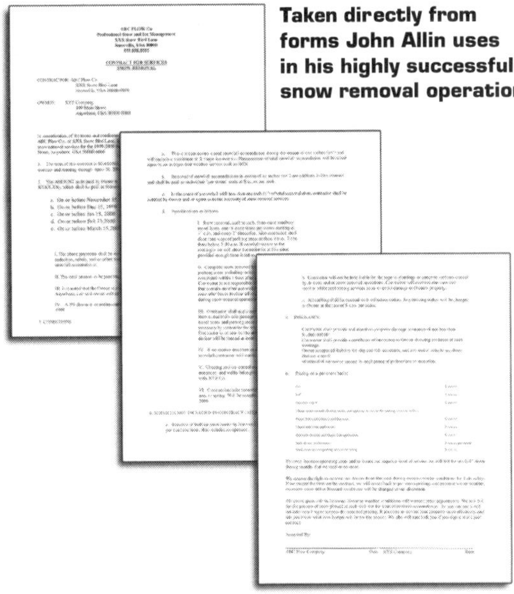

Taken directly from forms John Allin uses in his highly successful snow removal operation!

The Perfect Companion to *Managing Snow & Ice*

Contains all the forms and documents essential to snow removal.

More than 20 forms including:

- Sample contracts
- Subcontractor agreements
- Financial statements
- Snow response plans
- Marketing documents
- Time sheets
- Sales letter
- Job application
- And much more

$24.95

Lawn&Landscape MEDIA GROUP

FOUR EASY WAYS TO ORDER

1. CALL: 800-456-0707
2. MAIL to: Lawn & Landscape Media Group
   4012 Bridge Ave
   Cleveland, OH 44113
3. FAX to: 216-961-0364
4. VISIT: www.lawnandlandscape.com/store

Name ____________________

Title ____________________

Company ____________________

Address ____________________

City ____________________

State __________ Zip __________

Phone __________ Fax __________

E-Mail ____________________

	QTY	SUBTOTAL
Snow Removal Business Forms on CD-ROM – $24.95 each		$
Managing Snow & Ice Book – $49.95 each		$
COMBINATION – $59.95		$
Merchandise Total		$
Ohio residents – 7% sales tax		$
Shipping & Handling – Inside the U.S. – first item $5; additional items $1 each International – first item $10; additional items $3 each		$
TOTAL AMOUNT DUE		$

❒ Check enclosed
(Make payable to: Lawn & Landscape Media Group)

Charge my: ❒ American Express ❒ Visa ❒ Mastercard ❒ Discover/Novus

Card Number ____________________

Exp. Date ______/______